W9-BLB-360

BARRON'S

PAINLESS Pre-Algebra

Amy Stahl

© Copyright 2011 by Barron's Educational Series, Inc.

All rights reserved.
No part of this publication may be reproduced or distributed in any form or by any means without the written permission of the copyright owner.

All inquiries should be addressed to:
Barron's Educational Series, Inc.
250 Wireless Boulevard
Hauppauge, New York 11788
www.barronseduc.com

Library of Congress Control Number: 2010051392

ISBN: 978-0-7641-4588-9

Library of Congress Cataloging-in-Publication Data
Stahl, Amy.
Painless pre-algebra / Amy Stahl.
p. cm.
Includes bibliographical references and index.
ISBN 978-0-7641-4588-9 (alk. paper)
1. Mathematics. I. Title.
QA107.2.S73 2011
510—dc22 2010051392

PRINTED IN THE UNITED STATES OF AMERICA
9 8 7 6 5

CONTENTS

INTRODUCTION

Chapter One is about rational numbers. You will learn how to order, compare, add, subtract, multiply, and divide fractions painlessly. Fractions you say! No worries. With all of the tricks and hints in this chapter, you will be fearless of fractions.

Chapter Two reviews basic algebra. Names of properties are reviewed. Aunt Sally will help you remember the order of operations. We will also look at the difference between expressions and equations. By the end of this unit, you will be able to solve one-step equations painlessly!

Chapter Three shows you how to use proportions to solve many different types of problems. Whether you want to make cookies, read a map, or exchange money, all of this is done very easily with the use of proportions.

Chapter Four takes proportions a step further and explores percent problems. Did you know that shopping, using coupons, figuring the tax, and leaving a tip all involve painless proportions? Shopping is fun and involves proportions!

Chapter Five is about integers. Adding, subtracting, multiplying, and dividing, will all be as easy as pie once you learn the painless rules. Did you know pizza could help you remember what sign to use when you multiply or divide? So bring on the positive and negative numbers! You will be a pro once you are through this chapter.

Chapter Six deals with exponents and roots. Everyone likes learning shortcuts, and this will make multiplying numbers painless. Instead of writing $2 \times 2 \times 2 \times 2 \times 2$, wouldn't you rather write 2^5? This chapter will show you how to use exponents to rewrite a multiplication problem. Roots are easy, too, if you know $10 \times 10 = 100$. By the end of the chapter, you will know why the square root of 100 is 10. Exponents and roots, painless!

Chapter Seven is all about equations and inequalities. In this chapter, you will learn how to solve equations with parentheses and with variables on both sides. In chapter two, you learned how to solve basic equations. After this chapter, you will be excellent at solving multistep equations. You will also learn how to solve problems that involve greater than and less than!

In Chapter Eight, you will learn how to graph a line. All lines are straight. However, you will know the difference between horizontal lines and vertical lines and between lines with different slopes. If ski slopes are the only slopes you know, no worries after this chapter. You will be a pro at graphing lines using slope.

Chapter Nine shows you how to perform different operations with polynomials. What is a polynomial? You have used terms like $3x$ and $2x + 1$ when writing expressions. A polynomial is an expression with three or more terms. Don't worry, though. This chapter will make you a whiz at simplifying polynomials!

So have you figured out that this book is going to make you a stronger, more confident math whiz? Get your pencil ready—it's time to make pre-algebra painless!

Rational Numbers

RATIONAL NUMBERS

A rational number is a number that can be written as the fraction $\frac{a}{b}$, where $b \neq 0$.

The goal of simplifying a fraction is to make the numerator and denominator as simple as possible. In math, these are called **relatively prime numbers**. Relatively prime numbers have no common factors other than 1. For example, the factors of 3 are 3 and 1. The factors of 5 are 5 and 1. Since 3 and 5 have no factors in common other than 1, 3 and 5 are relatively prime.

A fraction can be simplified in two ways. Let's look at each method.

METHOD 1:

Simplify each fraction by dividing the numerator and the denominator by the same number until you cannot simplify any further. For example:

Simplify: $\frac{36}{48}$

$$\frac{36}{48}\frac{\div 2}{\div 2} = \frac{18}{24}\frac{\div 2}{\div 2} = \frac{9}{12}\frac{\div 3}{\div 3} = \frac{3}{4}$$

MATH TALK!

When simplifying fractions, it is VERY important to remember that whatever you do on the top of the fraction, you must also do on the bottom of the fraction.

Simplify: $\frac{8}{10}$

Both 8 and 10 can be divided by 2.
Divide 8 by 2 and 10 by 2!

$$\frac{8}{10}\frac{\div 2}{\div 2} = \frac{4}{5}$$

METHOD 2:

Divide the numerator and the denominator by the greatest common factor (GCF) . The greatest common factor is the largest number that will divide evenly into both numbers.

Simplify: $\frac{36}{48}$

To find the greatest common factor (GCF) of two numbers, list the factors of both numbers. You can make a chart like this:

36	1, 2, 3, 4, 6, 9, 12, 18, 36
48	1, 2, 3, 4, 6, 8, 12, 16, 24, 48

12 is the largest factor that will go into both 36 and 48. That means 36 and 48 can both be divided by 12.

Now simplify:

$$\frac{36}{48} = \frac{\div 12}{\div 12} = \frac{3}{4}$$

Let's try a few more.

Simplify: $-\frac{24}{36}$

METHOD 1:

$$-\frac{24}{36}\,\frac{\div 2}{\div 2} = -\frac{12}{18}\,\frac{\div 2}{\div 2} = -\frac{6}{9}\,\frac{\div 3}{\div 3} = -\frac{2}{3}$$

METHOD 2:

Find the GCF of 24 and 36.

24	1, 2, 3, 4, 6, 8, 12, 24
36	1, 2, 3, 4, 6, 9, 12, 18, 36

12 is the largest factor that will go into both 24 and 36. That means 24 and 36 can both be divided by 12.

Simplify: $\frac{15}{45}$

METHOD 1:

$$\frac{15}{45}\,\frac{\div 3}{\div 3} = \frac{5}{15}\,\frac{\div 5}{\div 5} = \frac{1}{3}$$

METHOD 2:

$$\frac{15}{45}\,\frac{\div 15}{\div 15} = \frac{1}{3}$$

MATH TALK!

When simplifying a fraction, whether you use method 1 or method 2 does not matter. Just make sure your fraction is in simplest form! That means it cannot be broken down any further.

BRAIN TICKLERS
Set # 1

Simplify.

1. $\frac{25}{75}$
2. $\frac{48}{72}$
3. $-\frac{36}{72}$
4. $\frac{9}{28}$
5. $-\frac{36}{80}$

(Answers are on page 23.)

WRITING FRACTIONS AS DECIMALS

Fractions can be changed to decimals using two methods. Method 1 involves dividing, and Method 2 involves using your calculator.

METHOD 1:

If you want to change a fraction like $\frac{3}{5}$ to a decimal, you have to divide. Set up the division problem just like you read the fraction. Since $\frac{3}{5}$ is read as "3 divided by 5," set up

$$\begin{array}{r} 0.6 \\ 5\overline{)3.00} \\ \underline{-30} \\ 0 \end{array}$$

→ Since the remainder is 0, this is a terminating decimal.

So the fraction $\frac{3}{5}$ is equivalent to the decimal 0.6.

Write $\frac{4}{9}$ as a decimal.

$$\begin{array}{r} 0.444 \\ 9\overline{)4.000} \\ \underline{-36}\downarrow \\ 40 \\ \underline{-36}\downarrow \\ 40 \\ \underline{-36} \\ 4 \end{array}$$

→ The pattern repeats. This is a repeating decimal.

The fraction $\frac{4}{9}$ is equivalent to the decimal 0.444 . . .

Caution—Major Mistake Territory!

When setting up a long division equation, the numerator must go underneath the division sign and the denominator must go outside of the division sign. Remember to set it up as you say it!

$$\frac{1}{2} = 1 \text{ divided by } 2 = 2\overline{)1}$$

METHOD 2

The easiest method to convert a fraction to a decimal, and a good way to check your long division, is to use a calculator.

How you read the fraction $\frac{3}{8}$, "3 divided by 8," is how you type it into the calculator, $3 \div 8$. Then just read the answer off of the calculator! $\frac{3}{8} = 0.375$

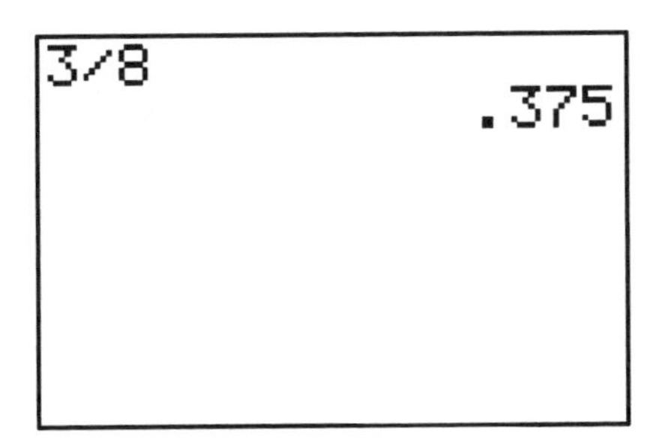

BRAIN TICKLERS
Set # 2

Write each fraction as a decimal.

1. $\frac{15}{20}$ 2. $\frac{5}{8}$ 3. $-\frac{5}{6}$ 4. $\frac{5}{4}$

(Answers are on page 23.)

WRITING DECIMALS AS FRACTIONS

To convert a decimal to a fraction, read the decimal using its place values. Let's review some place values.

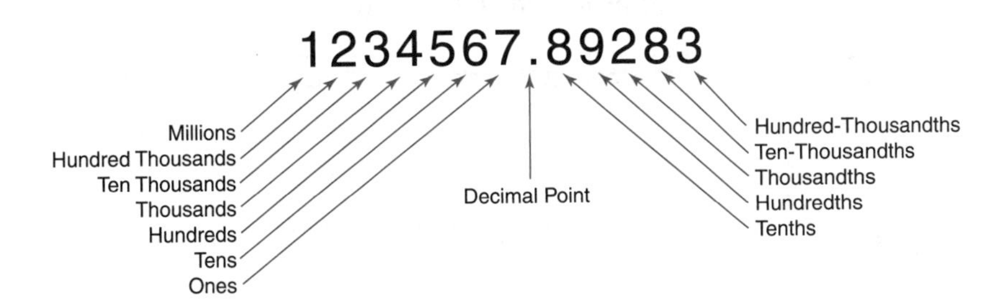

Write 0.75 as a fraction. The decimal is read as "75 one-hundredths." As a fraction, this is $\frac{75}{100}$. To simplify $\frac{75}{100}$, divide using the GCF method. The greatest common factor that will go into both 75 and 100 is 25. Divide both 75 and 100 by 25.

$$\frac{75}{100}\frac{\div 25}{\div 25} = \frac{3}{4}$$

Write 0.625 as a fraction. This decimal is read as "six hundred twenty-five thousandths." As a fraction, it is $\frac{625}{1000}$. Now simplify.

$$\frac{625}{1000}\frac{\div 25}{\div 25} = \frac{25}{40}\frac{\div 5}{\div 5} = \frac{5}{8}$$

Write 1.125 as a fraction. You read this decimal as "one and one hundred twenty-five thousandths." Remember that in math, "and" stands for the decimal point. So 1.125 as a fraction is $1\frac{125}{1000}$. The fraction $\frac{125}{1000}$ can be reduced, so the final fraction is $1\frac{1}{8}$.

SUM IT UP!

When converting decimals to fractions, read the decimal using its place values. Once you have the fraction, reduce to lowest terms

BRAIN TICKLERS
Set # 3

Write each decimal as a fraction.

1. 0.45 2. 3.5 3. –4.175 4. 0.04
5. 0.6

(Answers are on page 24.)

COMPARING AND ORDERING FRACTIONS AND DECIMALS

To compare and order rational numbers, you first have to write all of the numbers in the same form. A painless way to compare fractions is by finding a common denominator. You can use two methods to find the common denominator.

METHOD 1:

Multiply the denominators to find a common denominator.

Compare $\frac{5}{6}$ — $\frac{7}{8}$. Write >, <, or =.

Step 1: Multiply 6 and 8 to find a common denominator.
$6(8) = 48$

Step 2: Write the new fractions with the common denominator.
$\frac{5}{6} \cdot \frac{8}{8}$ — $\frac{7}{8} \cdot \frac{6}{6}$
$\frac{40}{48}$ — $\frac{42}{48}$

Step 3: Compare the fractions.
$\frac{40}{48} < \frac{42}{48}$ so $\frac{5}{6} < \frac{7}{8}$

METHOD 2:

Find the least common denominator.

Compare $\frac{5}{6}$ — $\frac{7}{8}$. Write >, <, or =.

Step 1: List the multiples of each denominator.
6: 6, 12, 18, 24, 30, . . .
8: 8, 16, 24, 32, . . .

The least common denominator is 24.

Step 2: Write the new fractions with the common denominator.

$\frac{5}{6} \cdot \frac{4}{4}$ ___ $\frac{7}{8} \cdot \frac{3}{3}$

$\frac{20}{24}$ ___ $\frac{21}{24}$

Step 3: Compare the fractions.

$\frac{20}{24} < \frac{21}{24}$ so $\frac{5}{6} < \frac{7}{8}$

MATH TALK!

Remember, the least common multiple (LCM) of two numbers is the smallest number, other than 0, that is a multiple of both numbers.

Find the LCM of 4 and 6.
4: 4, 8, 12, 16, . . .
6: 6, 12, 18, . . .
The LCM of 4 and 6 is 12.

If you are comparing and ordering numbers that are not all in the same format (all fractions, all decimals, or all percents), then you must put all of the numbers into the SAME format in order to compare them. The easiest thing to do is to turn every number into a decimal.

Step 1: Change every number into decimal form.

	0.75	$\frac{4}{5}$	0.5	$\frac{1}{10}$
	0.75	Divide 4 ÷ 5 on a calculator to get 0.8	0.5	Divide 1 ÷ 10 on a calculator to get 0.1
	0.75	0.80	0.50	0.10
Now order:	(#3)	(#4)	(#2)	(#1)

In order from least to greatest, the original numbers would read:
$\frac{1}{10}$, 0.5, 0.75, $\frac{4}{5}$.

MATH TALK!

It is easiest to compare numbers when they are all in the same format. Use your calculator to help you divide fractions to make them decimals.

MATH TALK!

When placing a list of numbers in order, be sure to list your final answer using the original numbers given to you. Writing them all as fractions is a good strategy when comparing them. However, the final answer should list the original numbers.

Actually numbering the problems, (#1), (#2), etc, will help you see the correct order.

BRAIN TICKLERS
Set # 4

Compare. Use >, <, or =.

1. $\frac{4}{7}$ — $\frac{3}{5}$
2. $\frac{6}{8}$ — $\frac{5}{9}$
3. $-\frac{3}{4}$ — $-\frac{4}{5}$
4. 0.36 — $\frac{2}{6}$
5. Order from least to greatest.

$0.3, \frac{1}{4}, 0.65, \frac{4}{5}, \frac{6}{10}$

(Answers are on page 24.)

ADDING AND SUBTRACTING FRACTIONS

Adding and subtracting fractions is easy. In order to add or subtract fractions, you need to have a common denominator. If the denominators are already the same, use the following rule.

> When the bottoms are the same, all you have to do is add the tops and then you're through.

If you want to add $\frac{5}{11} + \frac{4}{11}$, add the tops and then you're through. So $\frac{5}{11} + \frac{4}{11} = \frac{5+4}{11} = \frac{9}{11}$.

Use the same method for subtracting. To subtract $\frac{5}{6} - \frac{4}{6}$, subtract the tops and then you're through.

So $\frac{5}{6} - \frac{4}{6} = \frac{5-4}{6} = \frac{1}{6}$.

Before you can add or subtract fractions, you have to get a **common denominator**!

There are two methods to finding a common denominator. You can multiply one denominator by the other to create a common denominator. You can also find the least common denominator.

We used both of these methods when comparing and ordering fractions, but let's look at them again!

METHOD 1:
Multiply one denominator by another to get a common denominator

Add: $\frac{3}{5} + \frac{1}{6}$

Step 1: Multiply the denominators.
$5(6) = 30$

Step 2: To convert each fraction to an equivalent fraction, multiply by fractions equal to 1. To do this, multiply each fraction on the top and the bottom by the number needed to create 30.

For the denominator of 5, multiply the top and bottom by 6. For the denominator of 6, multiply the top and bottom by 5.

So: $\frac{3}{5}\left(\frac{6}{6}\right)+\frac{1}{6}\left(\frac{5}{5}\right)$

$\frac{18}{30}+\frac{5}{30}$

Step 3: Add the numerators, and keep the denominator.

$\frac{23}{30}$

Step 4: Simplify. Since $\frac{23}{30}$ cannot be reduced, it is already in simplest form.

MATH TALK!

Once you know the common denominator, multiply each fraction by the number needed to make that denominator.

For example, add $\frac{2}{3}+\frac{1}{7}$.

The common denominator is $3(7) = 21$.

So multiply the first fraction by $\frac{7}{7}$, and then multiply the second fraction by $\frac{3}{3}$.

$\frac{2}{3}\left(\frac{7}{7}\right)+\frac{1}{7}\left(\frac{3}{3}\right)$

$\frac{14}{21}+\frac{3}{21}=\frac{17}{21}$

METHOD 2:

Find the least common denominator.

Add: $\frac{3}{8} + \frac{1}{6}$

Step 1: List the multiples of each denominator.

8: 8, 16, 24, 32, . . .

6: 6, 12, 18, 24, . . .

The least common denominator is 24.

Step 2: Convert each fraction to an equivalent fraction by multiplying by fractions equal to 1. Since the common denominator is 24, multiply 8 by 3 to get 24, and multiply 6 by 4 to get 24. Remember, whatever you do on the top of the fraction, you must also do on the bottom!

$\frac{3}{8}\left(\frac{3}{3}\right) + \frac{1}{6}\left(\frac{4}{4}\right)$

$\frac{9}{24} + \frac{4}{24}$

Step 3: Add the numerators, and keep the denominators.

$\frac{13}{24}$

Step 4: Simplify if possible. $\frac{13}{24}$ already in lowest terms, so it cannot be further simplified.

BRAIN TICKLERS

Set # 5

Add or subtract. Simplify if possible.

1. $\frac{2}{9} + \frac{5}{9}$
2. $\frac{11}{15} - \frac{2}{15}$
3. $\frac{2}{5} + \frac{1}{3}$
4. $\frac{4}{7} - \frac{2}{8}$
5. $\frac{11}{12} - \frac{3}{4}$

(Answers are on page 24.)

MULTIPLYING FRACTIONS

Use these three easy steps to multiply fractions.

1. Multiply the numerators.
2. Multiply the denominators.
3. Simplify if possible.

A painless way to remember the rules is to use the following song.

> "Multiplying fractions, no big problem,
> multiply top by top and bottom by bottom!"

Multiply $\frac{3}{4}\left(\frac{5}{6}\right)$, and express the answer in simplest form.

1. Multiply the numerators: $3 \times 5 = 15$
2. Multiply the denominators: $4 \times 6 = 24$

So the answer is $\rightarrow \frac{3 \times 5}{4 \times 6} = \frac{15}{24}$.

3. Simplify the fraction if you can. 15 and 24 are both divisible by 3, so $\frac{15}{24} \div \frac{3}{3} = \frac{5}{8}$.

If you can do one problem, you can do them all! Multiply top by top and bottom by bottom!

Multiply: $\frac{3}{5} \cdot \frac{1}{4}$

1. Multiply the numerators: $3(1) = 3$
2. Multiply the denominators: $5(4) = 20$

The fraction is $\frac{3}{20}$. Since you cannot simplify 3 and 20, you are done!

Let's look at two more problems.

Multiply $5\left(\frac{4}{5}\right)$. Remember, any whole number is the same as the number divided by 1. So $5\left(\frac{4}{5}\right)$ is the same as $\frac{5}{1}\left(\frac{4}{5}\right)$. This makes it easy to multiply top by top and bottom by bottom! 5 times 4 = 20, and 5 times 1 = 5. So the fraction is $\frac{20}{5}$, which simplifies to 4.

Multiply $\frac{1}{4}\left(3\frac{2}{3}\right)$. Before we do this, we have to review mixed numbers.

MATH TALK!

To change a mixed number to an improper fraction multiply the denominator by the whole number, and then add the numerator. Write this number over the denominator.

$$2\frac{3}{8} = 8 \times 2 + 3 = \frac{19}{8}$$

Multiply: $\frac{1}{4}\left(3\frac{2}{3}\right)$

Step 1: Convert the improper fraction to a mixed number.

$$3\frac{2}{3} \rightarrow 3 \times 3 + 2 \rightarrow \frac{11}{3}$$

Step 2: Multiply top by top and bottom by bottom.

$$\frac{1}{4}\left(\frac{11}{3}\right) = \frac{11}{12}$$

Since $\frac{11}{12}$ does not simplify, we are done!

BRAIN TICKLERS
Set # 6

Multiply. Express in simplest form.

1. $\frac{4}{10}\left(\frac{3}{8}\right)$
2. $-\frac{4}{5}\left(\frac{5}{7}\right)$
3. $-4\left(\frac{2}{9}\right)$
4. $3\left(2\frac{1}{4}\right)$
5. $-\frac{4}{11}\left(-\frac{3}{5}\right)$

(Answers are on page 24.)

DIVIDING FRACTIONS

Use three simple steps to divide fractions. If you already know how to multiply, dividing is going to be easy as pie!

Step 1: Flip the second fraction over (take the reciprocal).

Step 2: Now multiply the two fractions.

Step 3: Simplify.

MATH TALK!

When you multiply a number by its reciprocal, the answer is 1.

The reciprocal of $\frac{3}{4}$ is $\frac{4}{3}$. Notice that $\frac{3}{4}\left(\frac{4}{3}\right) = 1$.

The reciprocal of 2 is $\frac{1}{2}$. Notice that $2\left(\frac{1}{2}\right) = 1$.

The reciprocal of $-\frac{5}{7}$ is $-\frac{7}{5}$. Notice that $-\frac{5}{7}\left(-\frac{7}{5}\right) = 1$.

The painless way is to use the following. "Dividing fractions is easy as pie, switch the second fraction and multiply!"

Divide: $\frac{4}{15} \div \frac{4}{5}$

Step 1: Flip the second fraction, so $\frac{4}{5}$ becomes $\frac{5}{4}$.

Step 2: Multiply the two fractions.

$$\frac{4}{15} \cdot \frac{5}{4} = \frac{20}{60}$$

Step 3: Simplify.

$$\frac{20}{60} = \frac{2}{6} = \frac{1}{3}$$

Caution—Major Mistake Territory!

When dividing fractions, always flip the second fraction because you are dividing by the second fraction.

When written in the form $\frac{\frac{1}{2}}{\frac{3}{4}}$, you can see that the second fraction is in the denominator. To remove this denominator, multiply both the numerator and denominator by the reciprocal of $\frac{3}{4}$, which is $\frac{4}{3}$. The problem would then become $\frac{1}{2} \cdot \frac{4}{3}$.

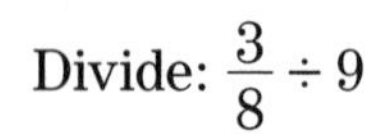

Divide: $\frac{3}{8} \div 9$

Step 1: Flip the second fraction. So $9 = \frac{9}{1}$ becomes $\frac{1}{9}$.

Step 2: Multiply the two fractions.

$$\frac{3}{8} \cdot \frac{1}{9} = \frac{3}{72}$$

Step 3: Simplify.

$$\frac{3}{72} = \frac{1}{24}$$

Let's try one more. Divide $-\frac{5}{12} \div \frac{10}{6}$. Remember the following chant.

"Dividing fractions is easy as pie,
switch the second fraction and multiply!"

$$-\frac{5}{12} \div \frac{10}{6} = -\frac{5}{12} \cdot \frac{6}{10} = -\frac{30}{120}$$

Now simplify $-\frac{30}{120}$.

$-\frac{30}{120} = -\frac{1}{4}$

BRAIN TICKLERS
Set # 7

Divide and simplify.

1. $\frac{2}{3} \div \frac{8}{7}$
2. $\frac{4}{9} \div \frac{12}{9}$
3. $-\frac{1}{2} \div -\frac{3}{4}$
4. $-\frac{3}{5} \div 6$

(Answers are on page 24.)

BRAIN TICKLERS—THE ANSWERS

Set # 1, page 5

1. $\frac{1}{3}$
2. $\frac{2}{3}$
3. $-\frac{1}{2}$
4. $\frac{9}{28}$
5. $-\frac{9}{20}$

Set # 2, page 8

1. 0.75
2. 0.625
3. –0.8333 . . .
4. 1.25

Set # 3, page 10

1. $\frac{9}{20}$
2. $3\frac{1}{2}$
3. $-4\frac{7}{40}$
4. $\frac{1}{25}$
5. $\frac{3}{5}$

Set # 4, page 14

1. $<$
2. $>$
3. $>$
4. $>$
5. $\frac{1}{4}$, 0.3, $\frac{6}{10}$, 0.65, $\frac{4}{5}$

Set # 5, page 17

1. $\frac{7}{9}$
2. $\frac{3}{5}$
3. $\frac{11}{15}$
4. $\frac{9}{28}$
5. $\frac{1}{6}$

Set # 6, page 20

1. $\frac{3}{20}$
2. $-\frac{4}{7}$
3. $-\frac{8}{9}$
4. $6\frac{3}{4}$
5. $\frac{12}{55}$

Set # 7, page 23

1. $\frac{7}{12}$
2. $\frac{1}{3}$
3. $\frac{2}{3}$
4. $-\frac{1}{10}$

CHAPTER TWO

Basic Algebra

PROPERTIES OF NUMBERS

You have a name. I have a name. Well, the rules for math also have a name. They are different types of properties! The following are the basic properties of the real numbers. The painless way is an easy explanation for each property. The examples will show you how they work.

NUMBER PROPERTIES

Property	The Painless Way	Examples
Commutative property of addition	Change order	$3 + 5 = 5 + 3$ $a + b = b + a$
Commutative property of multiplication	Change order	$3(5) = 5(3)$ $a(b) = b(a)$
Associative property of addition	Change grouping	$(3+4)+5=3+(4+5)$ $(a+b)+c=a+(b+c)$
Associative property of multiplication	Change grouping	$3 \cdot (4 \cdot 5) = (3 \cdot 4) \cdot 5$ $a \cdot (b \cdot c) = (a \cdot b) \cdot c$
Distributive property	Multiply each number inside the parentheses by the number in front of the parentheses	$2(3+4) = 2(3) + 2(4)$ $a(b + c) = ab + ac$
Additive identity property	What can you add to a number to get the original number? Add 0!	$5 + 0 = 5$ $a + 0 = a$
Multiplicative identity property	What can you multiply a number by to get the original number? Multiply by 1!	$5(1) = 5$ $a \cdot 1 = a$

Additive inverse property	Add a number's opposite to that number to get 0.	$5 + (-5) = 0$ $a + (-a) = 0$
Multiplicative inverse property	Multiply a number by its reciprocal to get 1.	$5 \cdot \frac{1}{5} = 1$ $a \cdot \frac{1}{a} = 1$
Multiplication property of zero	Any number times 0 equals 0.	$5(0) = 0$ $a \cdot 0 = 0$

Let's try a few, to see how easy naming these properties can be. Cover up the answer column, and see if you can guess the correct name, based on the clue!

EXAMPLES	CLUE	ANSWER
1. $9 + 3 = 3 + 9$	Order changed, addition	Commutative property of addition
2. $8+(4+3)=(8+4)+3$	Grouping changed, addition	Associative property of addition
3. $a(b + c) = ab + ac$	a multiplied to each	Distributive property
4. $14 + (-14) = 0$	Adding an opposite	Additive inverse property
5. $y + 0 = y$	Adding, getting back the identical number	Additive identity property
6. $-2(1) = -2$	Multiplying, getting back the identical number	Multiplicative identity property
7. $2\left(\frac{1}{2}\right) = 1$	Multiplying by a reciprocal	Multiplicative inverse property
8. $3(0) = 0$	Multiplying by zero	Multiplication property of zero

MATH TALK!

Remember these tricks!

CO → change order

Associative → who you associate with, group with, change groups

Distribute → hand out, multiply to each number

Additive identity → add 0 to get the identical number

Multiplicative identity → multiply by 1 to get the identical number

Additive inverse → add the opposite

Multiplicative inverse → multiply by the reciprocal

Multiplication property of zero → multiply by 0, you get 0!

BRAIN TICKLERS
Set # 8

For each example, name the property.

Commutative property of addition	Additive inverse property
Commutative property of multiplication	Multiplicative inverse property
Associative property of addition	Additive identity property

Associative property of multiplication	Multiplicative identity property
Multiplication property of zero	Distributive property

1. $a + c = c + a$
2. $a(bc) = (ab)c$
3. $4(5 + 7) = 4(5) + 4(7)$
4. $a + 0 = a$
5. $5 + (-5) = 0$
6. $(6 + 8) + 2 = 6 + (8 + 2)$
7. $6N = 6$
8. $AB = BA$
9. $\frac{4}{5} \cdot \frac{5}{4} = 1$
10. $5(0) = 0$

(Answers are on page 40.)

ORDER OF OPERATIONS

How would you find the answer to $(9 - 2) \times (4 + 3) + 3 - 12 \div 4 \div 3 + 16 - 1$ if you did not have a set of directions? Thank goodness for your Dear Aunt Sally!

To remember the order of operations, use the phrase "*P*lease *E*xcuse *M*y *D*ear *A*unt *S*ally." What does this mean in math?

P → simplify inside the parentheses
E → evaluate the exponents
MD → multiply and divide, left to right
AS → add and subtract, left to right

$3(4) + 5 - 1$	First multiply 3(4).
$12 + 5 - 1$	Add and subtract from left to right.
$17 - 1$	
16	

Caution—Major Mistake Territory!

3(4) is a multiplication problem. Remember in math that parentheses can stand for multiplication. Parentheses can also have a math problem inside them, like (3 + 4).

$5^2 - (16 + 2) \div 9$	First, add 16 + 2 in the parentheses.
$5^2 - 18 \div 9$	Evaluate the exponent, 5^2.
$25 - 18 \div 9$	Divide.
$25 - 2$	Subtract.
23	

$3(4^2) + 7$	First, evaluate the exponent 4^2.
$3(16) + 7$	Multiply 3(16).
$48 + 7$	Add.
55	

$3(5) - 4(3)$	Multiply 3 times 5.
$15 - 4(3)$	Multiply 4 times 3.
$15 - 12$	Subtract.
3	

BRAIN TICKLERS

Set # 9

1. $2 \times 5 + 15 \div 3$
2. $35 \div (15 - 8) + 1$
3. $3^2 - 12 \div 2 + 4 \times 2$
4. $3(4) + 2(1 + 3)^2 - 5$
5. $27 \div 3^2 - 4 + 8 \div 2$
6. $24 \div 6(2) - 2^2$

(Answers are on page 41.)

WRITING EXPRESSIONS

To write an expression, you need to know a few key words. Here is a chart to help you!

+	−	×	÷	EXPONENT
add	subtract	times	divided by	raised to
sum	difference	multiplied by	quotient	times itself
more than	less than	product	ratio of	squared
increased by	decreased by	of		cubed
exceeds	minus	twice		
in all	subtracted from	triple		
gain	reduced by	double		
total				
plus				

Exciting Examples

Write an expression for each phrase.

1. The sum of x and 5 — $x + 5$
2. The quotient of c and 7 — $c \div 6$ or $\frac{c}{6}$
3. e less than 7 — $7 - e$
4. The product of 6 and q — $6q$
5. u subtracted from 10 — $10 - u$
6. x cubed — x^3

Caution—Major Mistake Territory!

When writing expressions using "less than" and "subtracted from," the first number must come after the subtraction sign. For example,

5 less than x is NOT $5 - x$. It is $x - 5$.

Beware of subtraction!

BRAIN TICKLERS
Set # 10

Write an algebraic expression for each of the following.

1. 3 times x
2. 2 more than three times x
3. The difference of 4 and n
4. The sum of 3 and twice x
5. 8 increased by x
6. Twice the sum of x and 5
7. x to the second power
8. 8 times a number n
9. y divided by 12
10. 12 subtracted from x

(Answers are on page 41.)

EVALUATING EXPRESSIONS

You want to go to the movies with your friends. A movie ticket costs \$7. Let's see how much money you can spend on movie tickets. If you buy one ticket, you will spend \$7.

Look at the pattern: One ticket times \$7 = \$7
Two tickets times \$7 = \$14
Three tickets times \$7 = \$21

If each ticket costs \$7, this can be written as \$$7t$. This is called an algebraic expression.

An expression contains numbers, variables, and at least one operation. Examples of expressions are $2x$, $3y + 5$, and $xy^2 - 10w$.

Caution—Major Mistake Territory!

An expression has numbers, letters, and operations. An expression does NOT have an equals sign.

Algebraic expressions: $x + y$, $4ab + c$, $\frac{x}{2} + 3$

Not algebraic expressions: $x + 2 = 7$ $5 + 3 = y$

IMPORTANT VOCABULARY:

$$25n + 6$$

25 is the **coefficient**. A coefficient is the number multiplied by the variable

n is the **variable**. A variable is a letter that represents a number

6 is the **constant**. A constant has a value that does not change.

To **evaluate** an expression, substitute a given number for the variable and find the value of the expression.

Evaluate $2y + 1$ for $y = 4$. Plug in 4 for y.

$2y + 1$
$2(4) + 1$
$8 + 1$
9

Evaluate $4x + 3y$ for $x = 2$, $y = 5$. Plug in 2 for x and 3 for y.

$4x + 3y$
$4(2) + 3(5)$
$8 + 15$
23

Evaluate $2a^2 + 5b$ if $a = 3$, $b = 4$.

Substitute 3 for a and 4 for b. Remember to use order of operations.

$2a^2 + 5b$
$2(3^2) + 5(4)$
$2(9) + 5(4)$
$18 + 20$
38

Caution—Major Mistake Territory!

When evaluating, remember to use the order of operations: P, E, MD, AS. Exponents before multiplying!

$2(3^2) \rightarrow$ 3 is squared.

Then multiply by 2.

$2(9) = 18$

The expression $\$39.99 + \$0.25m$ represents a cell phone plan that costs \$39.99 a month plus \$0.25 for each additional minute, m, used over the plan. Find the cost of the phone bill for a month if you talk 4 minutes over the plan.

This is the same as $39.99 + 0.25m$ for $m = 4$. So plug in 4 for m!

$\$39.99 + \$0.25m$
$\$39.99 + \$0.25(4)$
$\$39.99 + \$1.00$
$\$40.99$

BRAIN TICKLERS
Set # 11

Evaluate each of the following for the given value(s).

1. $5x + 5$ for $x = 1$
2. $11 - 6m$ for $m = 4$
3. $3(z + 8)$ for $z = 3$
4. $3z - 3y$ for $z = 4, y = 2$
5. $4(2 + x) + 5$ for $x = 13$
6. Use the expression $1.8c + 32$ to convert the boiling point temperature of water from degrees Celsius to degrees Fahrenheit if the temperature is 10°C.

(Answers are on page 41.)

ONE-STEP EQUATIONS

An equation is a mathematical sentence that uses an equal sign to show that two expressions have the same value.

EQUATIONS	NOT EQUATIONS
$x + 5 = 12$	$2x + 5$
$x - 4 = 8$	$x - 4y$
$2x = 10$	
$\frac{x}{2} = 4$	

To solve an equation, you have to find the value of the variable (letter) that makes the sentence true. This answer is called the solution.

MATH TALK!

Addition and subtraction are inverse operations. They undo each other. To solve an equation, use the inverse operation to get the variable by itself.

To move to the opposite side of the equal sign, use the opposite sign!

Solve:

$$\begin{array}{rr} x + 5 = & 12 \\ -5 & -5 \\ \hline x = & 7 \end{array}$$

To undo adding 5, subtract 5 from both sides of the equation.

Check:
$x + 5 = 12$
$7 + 5 = 12$
$12 = 12$

$$\begin{array}{rr} x - 6 = & 13 \\ +6 & +6 \\ \hline x = & 19 \end{array}$$

To undo subtracting 6, add 6 to both sides of the equation.

Check:
$x - 6 = 13$
$19 - 6 = 13$
$13 = 13$

The same idea applies when you are multiplying or dividing in a problem.
To undo multiplication → divide.
To undo division → multiply.

$2x = 12$ This is read as "two times a number x equals 12." To undo multiplication by 2, divide by 2.

$$\frac{2x}{2} = \frac{12}{2}$$

$$x = 6$$

Check: $2x = 12$
$2(6) = 12$
$12 = 12$

$\frac{x}{3} = 4$ This is read as "a number x divided by 3 equals 4." To undo division by 3, multiply by 3.

$$\frac{x}{3} = 4$$

$$3 \cdot \frac{x}{3} = 4 \cdot 3$$

$$x = 12$$

Check: $\frac{x}{3} = 4$
$\frac{12}{3} = 4$
$4 = 4$

$\frac{x}{10} = 5$ To undo division by 10, multiply by 10.

$$\frac{x}{10} = 5$$

$$10 \cdot \frac{x}{10} = 5 \cdot 10$$

$$x = 50$$

Check: $\frac{x}{10} = 5$
$\frac{50}{10} = 5$
$5 = 5$

BRAIN TICKLERS
Set # 12

Solve each of the following.

1. $6x = 24$ 2. $\frac{x}{4} = 9$ 3. $22y = 44$
4. $12y = 144$ 5. $\frac{x}{8} = 6$ 6. $\frac{x}{11} = 10$

(Answers are on page 41.)

BRAIN TICKLERS—THE ANSWERS

Set # 8, page 29

1. Commutative property of addition
2. Associative property of multiplication
3. Distributive property
4. Additive identity property
5. Additive inverse property
6. Associative property of addition
7. Multiplicative identity property
8. Commutative property of multiplication
9. Multiplicative inverse property
10. Multiplication property of zero

Set # 9, page 32

1. 15
2. 6
3. 11
4. 39
5. 3
6. 4

Set # 10, page 34

1. $3x$
2. $2 + 3x$
3. $4 - n$
4. $3 + 2x$
5. $8 + x$ or $x + 8$
6. $2(x + 5)$
7. x^2
8. $8n$
9. $\frac{y}{12}$
10. $x - 12$

Set # 11, page 37

1. 10
2. –13
3. 33
4. 6
5. 65
6. 50°F

Set # 12, page 40

1. $x = 4$
2. $x = 36$
3. $y = 2$
4. $y = 12$
5. $x = 48$
6. $x = 110$

CHAPTER THREE
Ratios and Proportions
?
Roses
6 for
$13.99
Roses
8 for
$18.80

RATIO, RATE, UNIT RATE

A **ratio** is the comparison of two numbers, by division, that have the same unit.

You study 5 days out of 7 days. A recipe needs 4 cups of water for every 2 cups of flour.

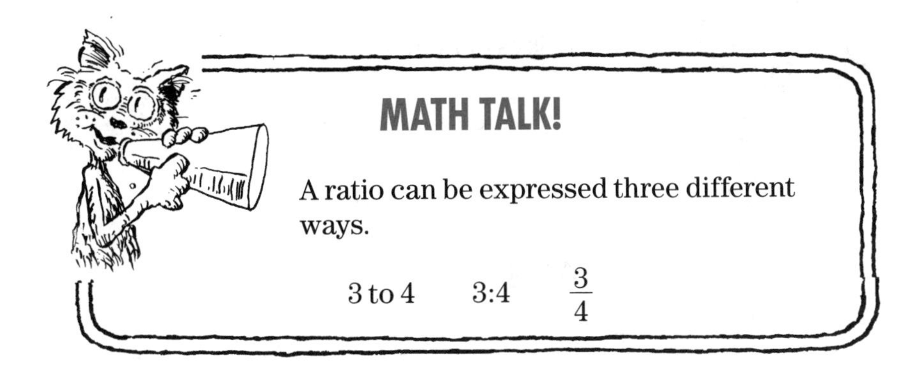

MATH TALK!

A ratio can be expressed three different ways.

3 to 4   3:4 $\frac{3}{4}$

A **rate** is the comparison of two numbers, by division, that have different units.

$30 spent on 10 gallons of gas   100 tickets sold in 4 days

A **unit rate** is a rate with a denominator of one.

$3.99 per gallon   55 miles per hour   20 students can sit in one row

Exciting Examples

Cover up the right column. Try to name each situation in the left column. Write *rate*, *ratio*, or *unit rate*. Then uncover the right column to check your answers.

1. Amy bought 3 bottles of water for $2.99.	rate
2. Reed spent $24 on 2 movies.	rate
3. It costs $0.25 per text message.	unit rate
4. The team won 9 games and lost 7.	ratio

BRAIN TICKLERS
Set # 13

Decide whether each is a ratio, rate, or unit rate.

1. 5 days per calendar year. ____
2. 12 students spent $60. ____
3. 15 boys to 12 girls. ____
4. A tutor charges $30 per hour. ____
5. Joe read 3 books in 7 days. ____

(Answers are on page 66.)

CALCULATING UNIT RATE

A unit rate is a rate with a denominator of one. Some examples of commonly used unit rates are miles per hour, cost per item, or earnings per week. For each unit rate, the first quantity is compared to 1 unit of the second quantity. So how do we calculate unit rate?

When given the rate $\frac{60 \text{ miles}}{3 \text{ hours}}$, simplify the ratio by dividing so the denominator becomes 1.

$\frac{60}{3} \frac{\div 3}{\div 3} = \frac{20}{1}$. Therefore the unit rate is $\frac{20 \text{ miles}}{1 \text{ hours}}$, or 20 mi/hr.

MATH TALK!

Rates and **unit rates** are comparisons of two numbers that have <u>different units</u>. It is very important to remember to label your units of measure when dealing with rates and unit rates!

Example 1

Chris made $32.50 in 4 hours. How much did Chris make in one hour?

Step 1: Write the rate as a fraction.

$$\frac{\$32.00}{4 \text{ hours}}$$

Step 2: Divide to find cost per hour.

$$\frac{\$32}{4 \text{ hours}} \frac{\div 4}{\div 4} = \frac{\$8}{1 \text{ hour}}$$

Chris made $8 per hour.

Example 2

Hudson is a starting player for his high school basketball team. He has scored 108 points in the last 6 games. Express this as a unit rate.

Step 1: Write the rate as a fraction.

$$\frac{108 \text{ points}}{6 \text{ games}}$$

Step 2: Divide to find the unit rate, points per game.

$$\frac{108 \text{ points}}{6 \text{ games}} \frac{\div 6}{\div 6} = \frac{18 \text{ points}}{1 \text{ game}}$$

Hudson scored 18 points per game. This is the unit rate!

Unit price is a unit rate used to compare price per item. Sometimes, instead of saying find the unit rate, a problem might say find the unit price.

Example 3

Find each unit price and tell which is the better buy.

$3.99 for 3 pounds of peanuts or $5.99 for 5 pounds of peanuts.

Step 1: Write each rate as a fraction.

$$\frac{\$3.99}{3 \text{ pounds}} \text{ or } \frac{\$5.99}{5 \text{ pounds}}$$

Step 2: Divide each rate to find the unit price for each item.

$$\frac{\$3.99}{3 \text{ pounds}} \frac{\div 3}{\div 3} = \frac{\$1.33}{1 \text{ pound}} \text{ or}$$

$$\frac{\$5.99}{5 \text{ pounds}} \frac{\div 5}{\div 5} = \frac{\$1.20}{1 \text{ pound}}$$

The better buy is the package of peanuts that cost $5.99 for 5 pounds.

Caution—Major Mistake Territory!

When finding unit price, or dividing any problem involving money, it is important to remember that money is always rounded to the nearest cent (hundredths). You have to remember to always check for rounding.

Example: $5.99/5 = $1.198 → since there is an 8 in the thousandths place, the nine will round up, making the answer $1.20.

$3.49/4 = $0.8725 → since there is a 2 in the thousandths place, the 7 does not round up. The answer is $0.87.

BRAIN TICKLERS
Set # 14

Find each unit rate.

1. A car travels 816 miles in 16 hours.
2. There are 260 students on 5 buses.
3. It costs $59.95 to buy 7 CDs.
4. Reed can wrap 52 presents in 4 hours. How many presents can Reed wrap per hour?
5. Which is the better buy? A bouquet of roses that cost $13.99 for 6 roses or a bouquet that costs $18.80 for 8 roses?

(Answers are on page 66.)

PROPORTIONS

A proportion is an equation that shows two equal ratios. A proportion can be written in three ways.

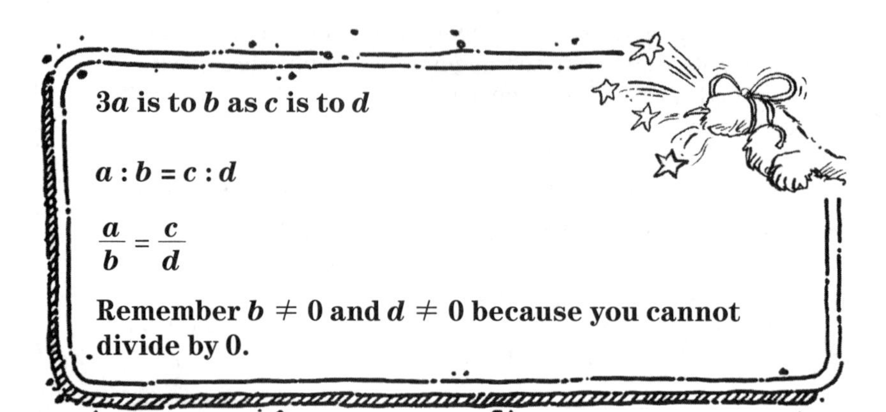

To solve a proportion, cross multiply. $\frac{a}{b} \times \frac{c}{d}$

$a(d) = b(c)$ → This is called the cross product.

We will use cross multiplying to help us solve for an unknown variable.

TYPE 1: DETERMINING IF RATIOS ARE EQUAL

Step 1: Set the ratios equal to each other.

Step 2: Cross multiply.

Step 3: If equal, it is a proportion.
If not equal, it is not a proportion.

Do the ratios $\frac{3}{4}$ and $\frac{9}{12}$ form a proportion?

$\frac{3}{4} \stackrel{?}{=} \frac{9}{12}$

$3 \times 12 \stackrel{?}{=} 9 \times 4$

$36 = 36$

Since the cross products are equal, the ratios are proportions.

Do the ratios $\frac{2}{5}$ and $\frac{8}{16}$ form a proportion?

$\frac{2}{5} \stackrel{?}{=} \frac{8}{16}$

$2 \times 16 \stackrel{?}{=} 8 \times 5$

$32 \neq 40$

Since the cross products are not equal, the ratios do not form a proportion.

TYPE 2: USING PROPORTIONS TO SOLVE PROBLEMS

Solve for x: $\frac{5}{8} = \frac{x}{48}$

$$5(48) = 8(x)$$

$$\frac{240}{8} = \frac{8x}{8}$$

$$x = 30$$

BRAIN TICKLERS
Set # 15

1. Do the ratios $\frac{2}{9}$ and $\frac{6}{12}$ form a proportion?

2. For the set of ratios, find the two that are proportional. $\frac{35}{26}, \frac{81}{39}, \frac{27}{13}$

3. Solve: $\frac{3}{12} = \frac{x}{60}$

4. Solve: $\frac{x}{9} = \frac{16}{6}$

(Answers are on page 66.)

PROPORTION WORD PROBLEMS

If you can set up and solve a proportion, you can solve word problems involving proportions.

TYPE 1: WORD PROBLEMS

You can use proportions to solve for an unknown quantity when comparing two things.

John can complete 14 math problems in 20 minutes. At that rate, how many math problems can he complete in 30 minutes?

Let x = number of math problems

Set up a proportion comparing math problems to minutes

Math problems to minutes $\rightarrow \frac{14}{20} = \frac{x}{30} \leftarrow$ math problems to minutes

$$14(30) = 20(x)$$

$$\frac{420}{20} = \frac{20x}{20}$$

$$x = 21$$

John can complete 21 problems in 30 minutes.

TYPE 2: MAP SCALE

You can use proportions to find distances on a map.

A map uses a scale of 1 cm : 50 miles. If two cities are 3.5 cm apart on a map, what is the actual distance in miles?

Let x = the actual distance in miles

Set up a proportion comparing cm to miles.

cm to miles $\rightarrow \frac{1}{50} = \frac{3.5}{x} \leftarrow$ cm to miles

$$x = 50(3.5)$$

$$x = 175$$

The actual distance is 175 miles.

TYPE 3: SCALE DRAWINGS

Scale drawings are used for models of houses, buildings, cars, and airplanes. The size of the model is in proportion to the actual size of the real object.

A scale drawing of a building is 1 in. = 10 ft. If the actual building is 220 feet tall, how tall is the model?

Let x = the height of the model

Set up a proportion comparing inches to feet.

$$\text{in. to ft.} \rightarrow \frac{1}{10} = \frac{x}{220} \leftarrow \text{in. to ft.}$$

$$1(220) = 10(x)$$

$$\frac{220}{10} = \frac{10x}{10}$$

The model is 22 feet tall.

BRAIN TICKLERS
Set # 16

1. Mrs. Smith's car uses 3 gallons of gas to go 96 miles. How many gallons of gas will her car use for a trip of 160 miles?
2. One inch on a map represents 140 miles. If the distance from Rochester to Manhattan is actually 328 miles, how many inches would this be on a map, to the nearest tenth?
3. On a scale drawing, a room with a length of 14 feet measures 2.5 inches. If the master bedroom is 3.5 inches on the scale drawing, how long is the actual room?
4. Emily can shovel 5 driveways in 40 minutes. How many driveways can she shovel in 60 minutes?

(Answers are on page 66.)

METRIC CONVERSIONS

If you visit Europe, you have to use the metric system of measurement. The basic units are meters (m), liters (L), and grams (g). Here are the basic metric units you should know.

BASIC / METRIC UNITS

Length	Weight	Liquid Capacity
1 cm = 10 mm	1 gram (g) = 1000 mg	1 L = 1000 mL
1 m = 100 cm	1 kg = 1000 g	
1 km = 1000 m		

We can also use proportions to convert metric measurements. Remember, a proportion is an equation that states two ratios are equal. Use the metric units table to help you set up the converting ratio. For example, 1 cm = 100 mm can be written as $\frac{1}{100}$.

Another way to remember the metric units is to use the saying "King Henry died by drinking chocolate milk." Each letter stands for a metric prefix.

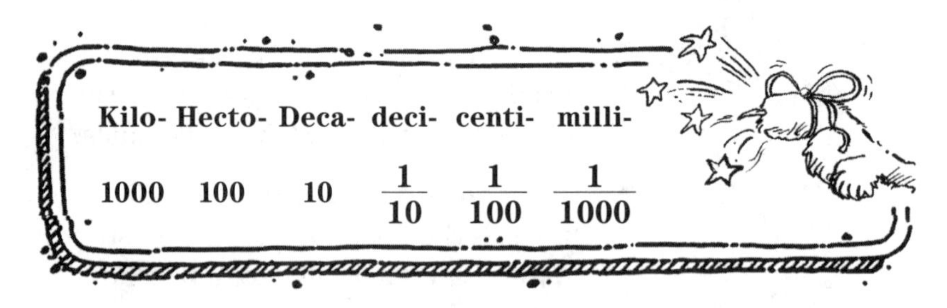

To multiply a metric unit by 10, 100, or 1000, move the decimal point to the right. To divide a metric unit by 10, 100, or 1000, move the decimal point to the left. So if you want to multiply by 100, move the decimal point 2 units to the right. If you want to divide by 1000, move the decimal point 3 units to the left.

Example 1

The mass of a sample of rocks is 1.24 kilograms. How many grams is this?

Let x = number of grams

Think of the problem as 1.24 kg = ____ g

Set up a proportion to compare kilograms to grams.

1.24 kg to x grams → $\frac{1.24}{x} = \frac{1}{1000}$ ← 1 kg = 1000 grams (Kilograms are larger than grams.)

Cross multiply: $1(x) = 1000(1.24)$

$x = 1240$

1.24 kilograms equal 1240 grams.

MATH TALK!

When using proportions to convert units, always set the larger unit as 1.

5 cm = ____ mm

cm mm	$\frac{5}{x} = \frac{1}{10}$	1 cm equals 10 mm (Centimeters are larger.)

Example 2

A soda bottle holds 2 liters of soda. How many milliliters does the bottle hold?

Let x = number of milliliters

Think of the problem as 2 L = ____ mm

Set up a proportion comparing liters to milliliters.

L to mL → $\frac{2}{x} = \frac{1}{1000}$ ← 1 L = 1000 mL (Liters are larger.)

Cross multiply: $1(x) = 1000(2)$
$x = 2000$
2 liters equals 2000 milliliters.

Example 3

How many centimeters are in 234 millimeters?

Let x = number of centimeters

Think of the problem as 234 mm = ____ cm

Set up a proportion comparing millimeters to centimeters.

mm to cm → $\frac{234}{x} = \frac{10}{1}$ ← 10 mm = 1 cm (Centimeters are larger.)

Cross multiply: $10(x) = 234(1)$

$$\frac{10x}{10} = \frac{234}{10}$$

$$x = 23.4$$

There are 23.4 centimeters in 234 millimeters.

BRAIN TICKLERS
Set # 17

Convert the units.

1. 89 km = ____ m
2. 5.8 cm = ____ m
3. How many kilometers are in 6,700 meters?
4. Convert 567 milligrams to grams.
5. A bottle holds 0.8 liters of water. How many milliliters is this?

(Answers are on page 66.)

CUSTOMARY CONVERSIONS

The United States uses customary units as our system of measurement. Here are the basic customary units of measure you should know.

BASIC CUSTOMARY UNITS

Length	Weight	Liquid Capacity
12 inches = 1 foot	16 ounces = 1 pound	8 ounces = 1 cup
3 feet = 1 yard	2000 pounds = 1 ton	2 cups = 1 pint
1760 yards = 1 mile		2 pints = 1 quart
5280 feet = 1 mile		4 quarts = 1 gallon

You can use proportions to solve customary conversion problems.

Example 1

The average weight of a baby at birth is 7 pounds. How many ounces is this?

Think of the problem as 7 lbs = ____ oz

You know that there are 16 ounces in a pound. Use this to set up a proportion.

Let x = number of ounces the baby is

pounds to ounces → $\frac{7}{x} = \frac{1}{16}$ ← 1 pound = 16 ounces

Cross multiply $x = 112$

A 7-pound baby weighs 112 ounces.

Example 2

The gasoline tank of a Toyota SUV holds 18 gallons of gas. How many quarts is this?

Think of the problem as 18 gallons = ____ quarts.

There are 4 quarts in a gallon. Set up a proportion.

Let x = number of quarts

gallons to quarts $\rightarrow \frac{18}{x} = \frac{1}{4} \leftarrow$ 1 gallon = 4 quarts

Cross multiply $72 = x$

18 gallons of gas is the same as 72 quarts.

Example 3

A football field is 100 yards long. How many feet is this?

Think of the problem as 100 yds = ____ ft.

There are 3 feet in a yard. Set up a proportion.

Let x = number of feet

yards to feet $\rightarrow \frac{100}{x} = \frac{1}{3} \leftarrow$ 3 feet = 1 yard

Cross multiply $300 = x$

A football field is 300 feet long.

Caution—Major Mistake Territory!

When setting up proportions, make sure the units are the same for both ratios. If you set up inches to feet on the left, the fraction on the right needs to be inches to feet.

For example, if 12 inches = 1 foot, how many inches equal 48 feet?

$\frac{\text{in.}}{\text{ft}}$ $\qquad \frac{x}{48 \text{ ft}} = \frac{12 \text{ in.}}{1 \text{ ft}}$

BRAIN TICKLERS
Set # 18

Convert each of the following.

1. 4 feet = ____ inches
2. 9 yards = ____ feet
3. You drove 17 miles to the mall and back. How many feet did you drive?
4. Chris used 18 gallons of gas for his truck this week. How many quarts did he use?
5. At the restaurant, you order a 24-ounce steak. How many pounds of steak are you ordering?

(Answers are on page 66.)

SIMILAR FIGURES

Two figures that have the same shape but not necessarily the same size are **similar**.

When two figures are similar, the ratio of the lengths of their corresponding sides is equal and the ratio of their corresponding angles is equal.

To determine if the triangles are similar, use a proportion to compare the corresponding sides.

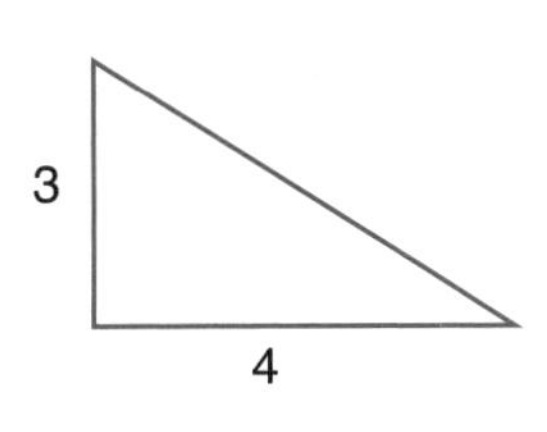

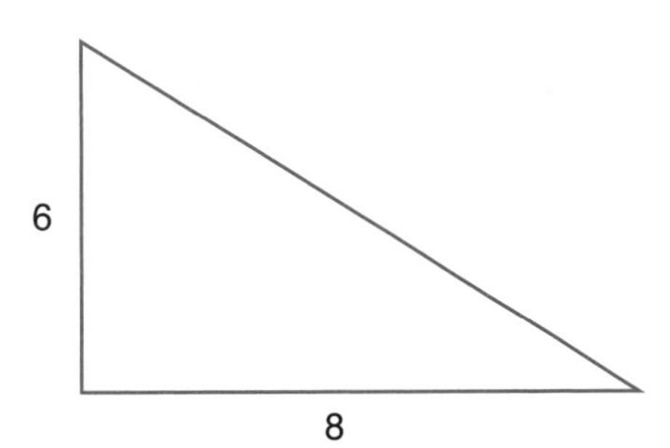

Set up a proportion comparing the side and bottom of the small triangle with the side and bottom of the larger triangle.

side to bottom of small triangle $\rightarrow \frac{3}{4} = \frac{6}{8} \leftarrow$ side to bottom of large triangle

Cross Multiply: $3(8) = 4(6)$
$24 = 24$

Since the two ratios are equal, the sides are proportional. This means the triangles are similar!

Example 1

A photo is 12 inches wide by 18 inches tall. If the width is scaled down to 9 inches, what is the height of the new photo?

Draw a picture to help you, and then set up a proportion comparing width to height (small side to large side).

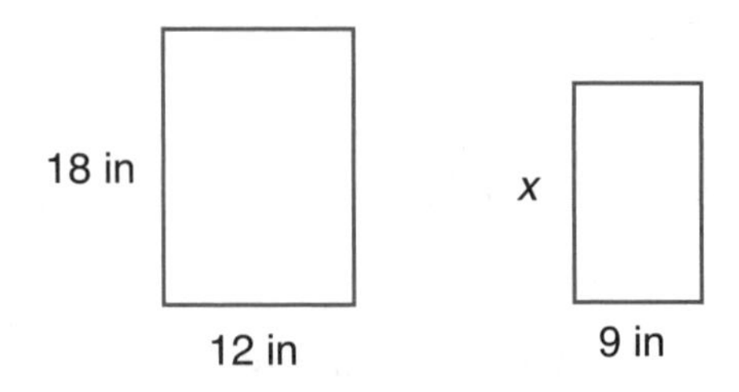

Let x = height of new (smaller) photo

height to width → $\frac{18}{12} = \frac{x}{9}$ ← height to width of new photo

Cross multiply $12x = 18(9)$

$$\frac{12x}{12} = \frac{162}{12}$$

$$x = 13.5$$

The height of the new photo is 13.5 inches.

MATH TALK!

Remember, a picture speaks a thousand words! To make problems easier to solve, draw a picture to help you! This is a *very* important tool for solving math problems.

Example 2

A building has a height of 192 meters. At the same time, a pole 4 meters high casts a shadow 16 meters long. What is the length of the shadow of the building?

Draw a picture to help you!

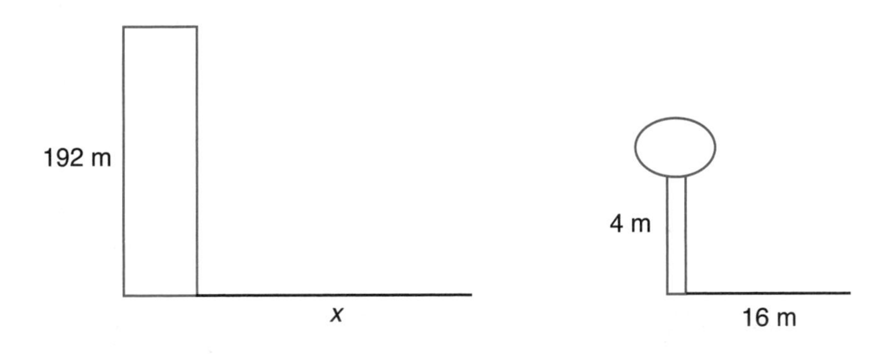

Let x = shadow of the building

Set up a proportion comparing the height and the shadow.

height to shadow of building $\rightarrow \dfrac{192}{x} = \dfrac{4}{16} \leftarrow$ height to shadow of pole

Cross Multiply:

$$\begin{aligned} 192(16) &= 4x \\ \frac{3072}{4} &= \frac{4x}{4} \\ 768 &= x \end{aligned}$$

The shadow of the building is 768 feet long.

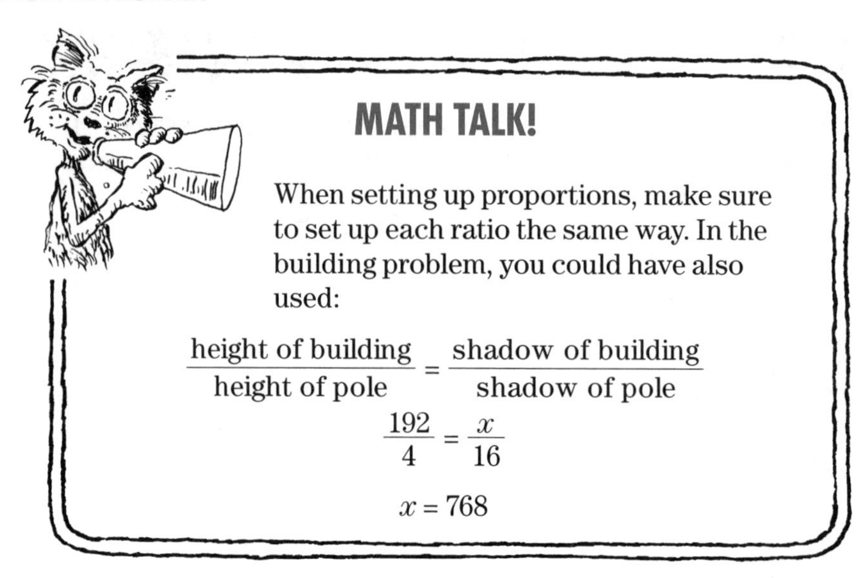

MATH TALK!

When setting up proportions, make sure to set up each ratio the same way. In the building problem, you could have also used:

$$\frac{\text{height of building}}{\text{height of pole}} = \frac{\text{shadow of building}}{\text{shadow of pole}}$$

$$\frac{192}{4} = \frac{x}{16}$$

$$x = 768$$

Example 3

A farmer wants to make a bridge that goes across the stream on his property. The diagram below shows the measurements that the farmer knows. The triangles drawn are similar. How wide is the stream where the farmer wants to build the bridge?

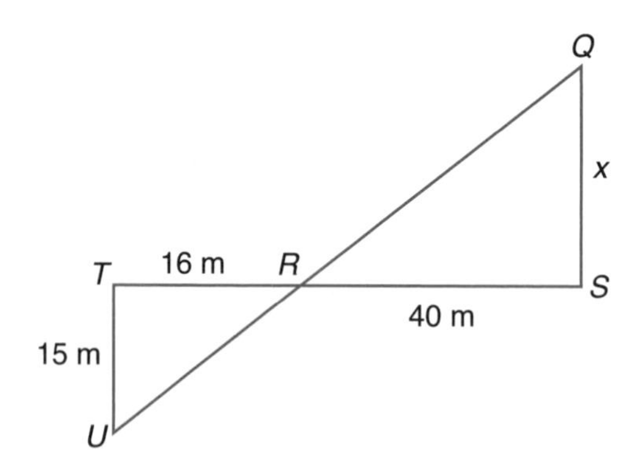

Since triangle RQS and RUT are similar, we can set up a proportion to solve for the missing side, x.

Again, it is important to make sure the ratios are set up the same way, especially if one triangle is flipped over, like in this example.

Set up a proportion comparing the side of the triangle to the bottom of the triangle.

Side to bottom of small triangle → $\frac{15}{16} = \frac{x}{40}$ ← Side to bottom of large triangle

Cross Multiply: $\frac{16x}{16} = \frac{600}{16}$

$x = 37.5$

The stream is 37.5 m wide.

BRAIN TICKLERS
Set # 19

1. Chris is 6 feet tall and casts a shadow that is 15 feet long. At the same time, a tree casts a shadow that is 135 feet long. How tall is the tree?

2. Chris wants to know the width of the pond on his farm. He drew the following diagram and labeled it with the measurements he knew. If the triangles drawn are similar, how wide is the pond?

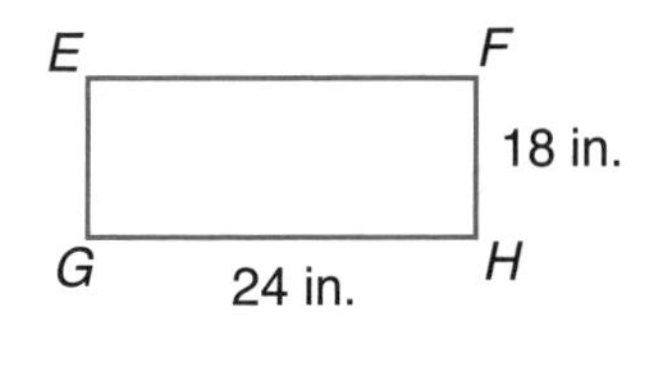

3. Find missing side *CD* if rectangle *ABCD* is similar to rectangle *EFGH*.

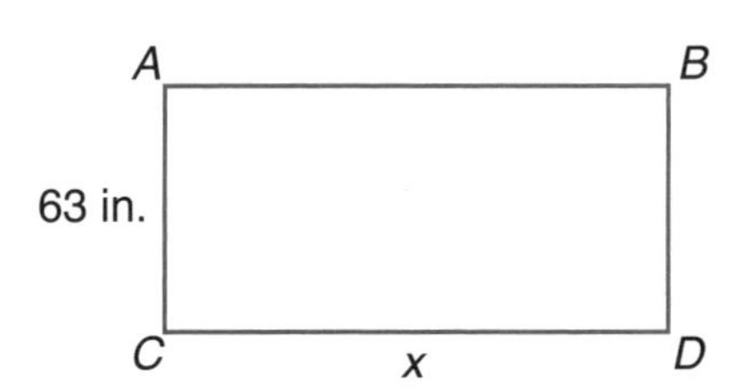

(Answers are on page 66.)

BRAIN TICKLERS—THE ANSWERS

Set # 13, page 46

1. Rate
2. Rate
3. Ratio
4. Unit rate
5. Rate

Set # 14, page 49

1. 136 miles/hour
2. 52 students/bus
3. $8.56/CD
4. 13 presents/hour
5. $13.99 for 6 roses

Set # 15, page 51

1. No
2. $\frac{81}{39}$ and $\frac{27}{13}$
3. $x = 15$
4. $x = 24$

Set # 16, page 53

1. 5 gallons
2. 2.3 inches
3. 19.6 feet
4. 7.5 driveways

Set # 17, page 57

1. 89,000
2. 580 m
3. 6.7 km
4. 0.567 g
5. 800 mL

Set # 18, page 60

1. 48 inches
2. 27 feet
3. 89,760 feet
4. 72 quarts
5. 1.5 pounds

Set # 19, page 65

1. 54 feet tall
2. 36 yards
3. 84 inches

CHAPTER FOUR
Percents
THAT'S NOT FAIR! YOU GUYS ALL GOT 25%, AND I ONLY GOT 1/4!

FRACTION, DECIMAL, AND PERCENT CONVERSIONS

Three out of four people own a cell phone. Three out of four is a ratio and can be expressed in three different ways.

A **ratio** is a comparison of two numbers and can be written as a fraction. → $\frac{3}{4}$

The ratio can be divided and written as a decimal. → $3 \div 4 = 0.75$

The decimal 0.75, read as "75 hundredths," can be written as a fraction. → $\frac{75}{100}$

Since percent means "out of 100," $\frac{75}{100}$ can be also written as 75%.

Here are some steps to make converting between decimals, fractions, and percents painless.

MAKING CONVERSIONS

Type of Problem	How to Convert	Example
Decimal to percent	Move decimal point two places to the right because you are multiplying by 100, then add the percent (%) sign	.25 = 25% 0.107 = 10.7% 0.005 = 0.5 % 1.5 = 150%
Percent to decimal	Drop percent sign and move the decimal point two places to left because you are dividing by 100.	30% = 0.30 5% = 0.05 120% = 1.2 0.5% = 0.005
Fraction to decimal	Use your calculator to divide the numerator by the denominator	$\frac{3}{4} = 0.75$ $\frac{1}{8} = 0.125$
Percent to fraction	Write number over 100, reduce fraction	$40\% = \frac{40}{100} = \frac{2}{5}$ $5\% = \frac{5}{100} = \frac{1}{20}$

MATH TALK!

Since percent is always out of 100, try to convert the fraction to an equivalent with a denominator of 100.

$$\frac{1}{5} = \frac{20}{100} \text{ so } \frac{1}{5} = 20\%$$

To make a percent a decimal, move the decimal point two places to the left and remove the percent sign.

Let's Practice

Fill in the chart. Check your answers below.

FRACTION	DECIMAL	PERCENT
$\frac{2}{5}$		
	0.65	
		5%
	0.7	
		30%
$\frac{7}{25}$		

ANSWERS:

FRACTION	DECIMAL	PERCENT
$\frac{2}{5}$	0.4	40%
$\frac{13}{20}$	0.65	65%
$\frac{1}{20}$	0.05	5%
$\frac{7}{10}$	0.7	70%
$\frac{3}{10}$	0.3	30%
$\frac{7}{25}$	0.28	28%

BRAIN TICKLERS
Set # 20

Complete the chart.

FRACTION	DECIMAL	PERCENT
	0.08	
$\frac{1}{20}$		
		14%
	0.72	
		4%
$\frac{3}{4}$		

(Answers are on page 87.)

PERCENT OF A NUMBER

You invited 50 friends to a party, and 70% of them said that they could come. How many friends are coming to the party?

In order to answer the question, you have to know how to find the percent of a number. You already know that percent is out of 100, so you have $\frac{70}{100}$. Well, if part of your 50 friends are coming, isn't that a fraction? You could say $\frac{x}{50}$ for the number of friends coming out of 50. We can use a proportion to figure this out!

$$\frac{\%}{100} = \frac{\text{is (part)}}{\text{of (whole)}}$$

Let's see how many friends are coming to the party.

70% is out of 100 $\rightarrow \frac{70}{100} = \frac{x}{50} \leftarrow x$ friends out of 50

Cross multiply:
$$70(50) = 100x$$
$$\frac{3500}{100} = \frac{100x}{100}$$
$$35 = x$$

35 friends are coming to the party!

Let's try a few more:

Example 1

\$150 is what percent of \$2400?

This time we are looking for the percent, so let x = the percent.

Percent over 100 $\rightarrow \frac{x}{100} = \frac{150}{2400} \leftarrow$ 150 is part of 2400

Cross multiply:
$$2400(x) = 150(100)$$
$$\frac{2400x}{2400} = \frac{15000}{2400}$$
$$x = 6.25$$

Remember you were looking for a percent, so 150 is 6.25% of 2400.

Example 2

30% of what number is $2400?

This time we are looking for the "of" or the total.

Percent over 100 → $\frac{30}{100} = \frac{2400}{x}$ ← 2400 is part of what number x?

Cross multiply:

$$30(x) = 2400(100)$$
$$\frac{30x}{30} = \frac{240000}{30}$$
$$x = 8000$$

30% of 8000 is 2400.

MATH TALK!

When dealing with percent problems, remember these points.
Percent is always out of 100.
The part is always out of the total.
The word "is" is part.
The word "of" is whole.

BRAIN TICKLERS
Set # 21

1. What is 125% of 80?
2. 35% of what number is 56?
3. What percent of 60 is 12?
4. What number is 40% of 1600?

(Answers are on page 87.)

DISCOUNT AND SALE PRICE

Shopping for new clothes, music, or books is fun! To be a good shopper, you need to understand sales tax and discounts.

The sale price is the discounted price of an item.

To find the sale price of an item, use these two easy steps.

Step 1: Find the amount of the discount.

$$\frac{\%}{100} = \frac{\text{is (part)}}{\text{of (whole)}}$$

Step 2: Subtract the amount of the discount from the original price.

Example 1

A stereo is on sale for 20% off the regular price of $389. What is the discount?

Step 1: Find the amount of the discount.

$$\frac{20}{100} = \frac{x}{389}$$

Cross multiply $100(x) = (20)(389)$

$$\frac{100x}{100} = \frac{7780}{100}$$

$$x = 77.80$$

The discount is $77.80.

Since the question asked for the discount, there is no second step.

Example 2

An iPod that regularly sells for $79.99 is on sale for 25% off this week only. Find the sale price.

Step 1: Find the amount of the discount.

$$\frac{25}{100} = \frac{x}{79.99}$$

Cross multiply $100(x) = (25)(79.99)$

$$\frac{100x}{100} = \frac{1999.75}{100}$$

$x = 19.9975$

Since this is money, round 19.997 to $20.00.

The discount is $20.00 off.

Step 2: Subtract the amount of the discount from the original price.
$79.99 – $20.00 = $59.99.

The sale price of the iPod is $59.99.

Caution—Major Mistake Territory!

Money is always rounded to the nearest cent unless otherwise stated. Since cents is out of 100, look at the 3rd number to the right of the decimal to see if you round up or down.

$\$20.05\underline{6} \rightarrow \$20.06$

$\$19.46\underline{2} \rightarrow \$19.46$

Example 3

A digital camera that normally retails for $495.95 is on sale for 33% off. What is the sale price of the camera?

Step 1: Find the amount of the discount.

$$\frac{33}{100} = \frac{x}{495.95}$$

$$\frac{100x}{100} = \frac{16366.35}{100}$$
$$x = 163.6635$$

The discount is $163.66 off.

Step 2: Subtract the amount of the discount from the original price.
$495.95 – $163.66 = $332.29

The sale price of the camera is $332.29.

BRAIN TICKLERS
Set # 22

1. Flat-screen TVs are on sale for 30% off this week. If a flat-screen TV is regularly priced at $899, what is the amount of the discount?

2. A new CD player regularly costs $49.99. This week it is on sale for 15% off the regular price. Find the sale price of the CD player.

3. Cameron has two coupons, one for $10 off an item and one for 20% off an item. If she buys a sweater for $39.99, which coupon would give her the best sale price?

4. Tracy has $50 to spend on a pair of shoes. Which pair of shoes can she afford to buy?

Pair of Shoes #1	Pair of Shoes #2
$59.99 regular price	$64.99 regular price
10% off	25% off

(Answers are on page 87.)

TAX

Sales tax is added to the price of an item or a service. Sales tax is a percent of the purchase price. A sales tax of 7% means that the item purchased will have an additional 7% added to its total cost.

To find the sales tax, use these two easy steps.

Step 1: Find the amount of tax using $\frac{\%}{100} = \frac{\text{is (part)}}{\text{of (whole)}}$.

Step 2: Add the tax to the original price.
Think tax → "t" looks like a plus.

Reed purchased a sweatshirt that costs $39.99. The tax rate at the mall is 7%. Find the amount of the sales tax.

Step 1: Find the amount of tax Reed pays.

$$\frac{\%}{100} = \frac{\text{is (part)}}{\text{of (whole)}}$$

$$\frac{7}{100} = \frac{x}{39.99}$$

$$\frac{100x}{100} = \frac{279.93}{100}$$

$$x = 2.79\underline{9}3$$

$$x = \$2.8\overline{0} \text{ tax}$$

The tax is $2.80.

Since the question asked for the amount of tax, do not do step 2.

Exciting Example

Mrs. Smith bought a new flat-screen TV. The retail price of the TV is $1499. If the sales tax is 8.25%, what is the final cost of the TV?

Step 1: Find the amount of tax.

$$\frac{8.25}{100} = \frac{x}{1499}$$

$$\frac{100x}{100} = \frac{12366.75}{100}$$

$$x = 123.66\underline{7}5$$

The tax is $123.67.

Step 2: Add the tax to the original price
$1499 + $123.67 = $1622.67

The final cost is $1622.67.

BRAIN TICKLERS
Set # 23

1. Your book purchase comes to $48.78. If the sales tax rate is 8%, find the amount you will pay in tax.

2. The tax rate in town is 7.5%. If you buy 2 sweaters for $19.99 each and a CD for $13.99, how much sales tax do you owe?

3. A new cell phone costs $129.99. The tax rate in Ontario County is 7.25%. Find the total cost of buying a cell phone in Ontario County.

4. John wants to buy 2 new video games with the money he received for his birthday. He has a total of $110 to spend. Each video game costs $49.99 plus tax. Does he have enough money to buy both games if the tax rate is 8%?

(Answers are on page 87.)

INTEREST

If you put money into a savings account, you earn interest. If you borrow money to buy a car, you pay interest. Since money is saved or borrowed over a period time, an interest rate is applied.

The **interest rate** is a percent over a fixed period of time.

The **principal** is the initial amount borrowed or invested.

Simple interest is the interest that is paid on only the initial amount.

To figure out how much interest is paid or borrowed, use the formula $I = Prt$.

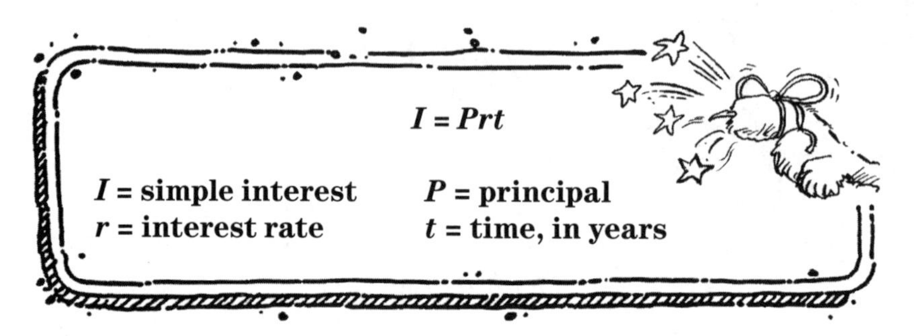

To find the interest, simply substitute the given values for each part of the formula.

A Few Examples

Find the interest on $345 at 4% per year for 3 years.

$I = Prt$

$I = 345(0.04)(3)$

$I = \$41.40$

Find the interest on $780 at 3.5% for 4 years.

$I = Prt$

$I = 780(0.035)(4)$

$I = \$109.20$

Find the interest on $800 at 11% for 3 months.

$I = Prt$

$I = 800(0.11)(3/12)$

$I = \$22$

Caution—Major Mistake Territory!

When writing percents as decimals, move the decimal two places to the left!

To use the formula $I = Prt$, time must be in years. If months are given, write the months as a fraction out of 12.

To find total amount including the interest, follow these two painless steps.

Step 1: Find the interest using $I = Prt$.

Step 2: Add the interest to the initial amount.

Example 1

$600 was deposited into an account that earns 2% interest. If no money was deposited or withdrawn from the account, how much will be in the account after 3 years?

Step 1: Find the interest.
$I = Prt$
$I = 600(0.02)(3)$
$I = \$36$
Interest = $36

Step 2: Add the interest to the initial amount.
$36 + $600 = $636

Total in account = $636

Example 2

You borrowed \$1000 to buy a car. How much money will you owe the bank if you borrow the money for 6 months at a rate of 6.5%?

Step 1: Find the interest.
$I = Prt$
$I = 1000(0.065)(6/12)$
$I = 32.5$

Step 2: Add the interest to the initial amount.
\$1000 + \$32.50 = \$1032.50

You will owe: \$1032.50

BRAIN TICKLERS
Set # 24

1. If \$620 is borrowed at 5.5% interest for 2 years, how much interest will be paid? How much money will be paid back?

2. If \$500 is borrowed at 6% interest for 3 months, how much interest will be paid? How much money will be paid back?

3. A bank offers an annual simple interest of 6% on home improvement loans. How much would Chris owe if he borrowed \$24,000 over a period of 2 years?

4. Kelly deposited \$425 in a savings account. How much would she have in the account after 3 years if the interest rate at the bank is 2%?

(Answers are on page 87.)

PERCENT INCREASE AND DECREASE

Prices rise and fall. Temperatures go up and down. People gain and lose weight. This change can be described by using a percent.

Percent change is the ratio of the amount of change to the original amount.

$$\frac{\text{change}}{\text{original}}$$

A percent increase describes how much the original amount increases. A percent decrease describes how much the original amount decreases. Here are examples of percent increase and decrease:

PERCENT INCREASE	PERCENT DECREASE
From 10 to 15 is an increase.	From 20 to 10 is a decrease.
Class size went from 15 to 22.	A $50 dress was marked down to $30.

To find the percent increase or decrease, follow these steps:

Step 1: Subtract the new amount from the original amount to find the change.

Step 2: Set up a proportion and solve. $\frac{\text{change}}{\text{original}} = \frac{x}{100}$

Example 1

Find the percent increase of $20 to $45.

Step 1: Subtract the new amount from the original.

$45 - 20 = 15$

Step 2: Set up a proportion and solve.

$$\frac{\text{change}}{\text{original}} = \frac{x}{100}$$

$$\frac{15}{20} = \frac{x}{100}$$

$$\frac{20x}{20} = \frac{1500}{20}$$

$$x = 75$$

The increase from $20 to $45 is 75%.

Example 2

Find the percent decrease of $50 to $40.

Step 1: Subtract the new amount from the original.

$50 - 40 = 10$

Step 2: Set up a proportion and solve.

$$\frac{\text{change}}{\text{original}} = \frac{x}{100}$$

$$\frac{10}{50} = \frac{x}{100}$$

$$\frac{50x}{50} = \frac{1000}{50}$$

$$x = 20$$

The decrease from $50 to $40 is 20%.

Example 3

Tot Toys buys bikes for $65 each and sells them for $80 each. Find the percent increase on the bike to the nearest percent.

Step 1: The original price was $65. The new price is $80. Subtract the two.

$\$80 - \$65 = \$15$

There is a $15 increase in price.

Step 2: Set up a proportion and solve.

$$\frac{\text{change}}{\text{original}} = \frac{x}{100}$$

$$\frac{15}{65} = \frac{x}{100}$$

$$65(x) = 15(100)$$

$$\frac{65x}{65} = \frac{1500}{65}$$

$$x = 23.076\ldots$$

The increase from $65 to $80 is approximately 23%.

Caution—Major Mistake Territory!

The change must always be divided by the original amount.

Example 4

John bought a video game system for $225. The original price of the system was $350. What was the percent decrease in price to the nearest percent?

Step 1: The original price was $350 and the new price is $225. Subtract the two.

$\$350 - \$225 = \$125$

The decrease in price is $125.

Step 2: Set up a proportion and solve.

$$\frac{\text{change}}{\text{original}} = \frac{x}{100}$$

$$\frac{125}{350} = \frac{x}{100}$$

$$350(x) = 125(100)$$

$$\frac{350x}{350} = \frac{12500}{350}$$

$$x = 35.71\ldots$$

The price has decreased by approximately 36%.

BRAIN TICKLERS
Set # 25

1. Find the percent decrease of $35 to $24.95 to the nearest percent.

2. Find the percent increase of $195 to $350 to the nearest percent.

3. A winter coat that usually sells for $425 is on sale for $369. What is the percent decrease to the nearest percent?

4. On the first day of school, Mrs. Linehan had 19 students in her math class. Now she has 25. Find the percent increase to the nearest percent.

5. Mr. Smith owns a bike store and typically marks up the bikes 20% over the price he pays for the bikes. How much would he charge for a bike that cost him $69.95?

(Answers are on page 87.)

BRAIN TICKLERS—THE ANSWERS

Set # 20, page 71

FRACTION	DECIMAL	PERCENT
$\frac{2}{25}$	0.08	8%
$\frac{1}{20}$	0.05	5%
$\frac{7}{50}$	0.14	14%
$\frac{18}{25}$	0.72	72%
$\frac{1}{25}$	0.04	4%
$\frac{3}{4}$	0.75	75%

Set # 21, page 74

1. 100
2. 160
3. 20%
4. 640

Set # 22, page 77

1. $269.70
2. $42.49
3. $10 off coupon
4. Pair #2 ($48.74)

Set # 23, page 79

1. $3.90 tax
2. $4.05 tax
3. $131.41 total
4. Yes (total = $107.98)

Set # 24, page 82

1. Interest = $68.20; Total = $688.20
2. Interest = $7.50; Total = $507.50
3. $2880
4. $450.50

Set # 25, page 86

1. 29% decrease
2. 79% increase
3. 13% increase
4. 32% increase
5. $83.94

CHAPTER FIVE

Integers

GRAPHING POINTS

Have you ever had to find a doctor's office in a building? If someone were to give you directions that person might say, "Take a right down the hallway to the stairs. Walk up 3 flights of stairs, and take a left." Graphing points uses the same idea. You always start from a center point and then move either left or right and move either up or down. To help us visualize positive and negative numbers, we can graph them on a coordinate system.

The **coordinate system** is formed by the intersection of two lines, the x-axis and the y-axis. To help you remember which is which, think "y in the sky" for the y-axis. The x-axis goes left and right, and the y-axis goes up and down.

The four regions created by the axes are called the **quadrants**. The point located at the center of the axes is called the **origin**. The coordinates of the origin are (0,0).

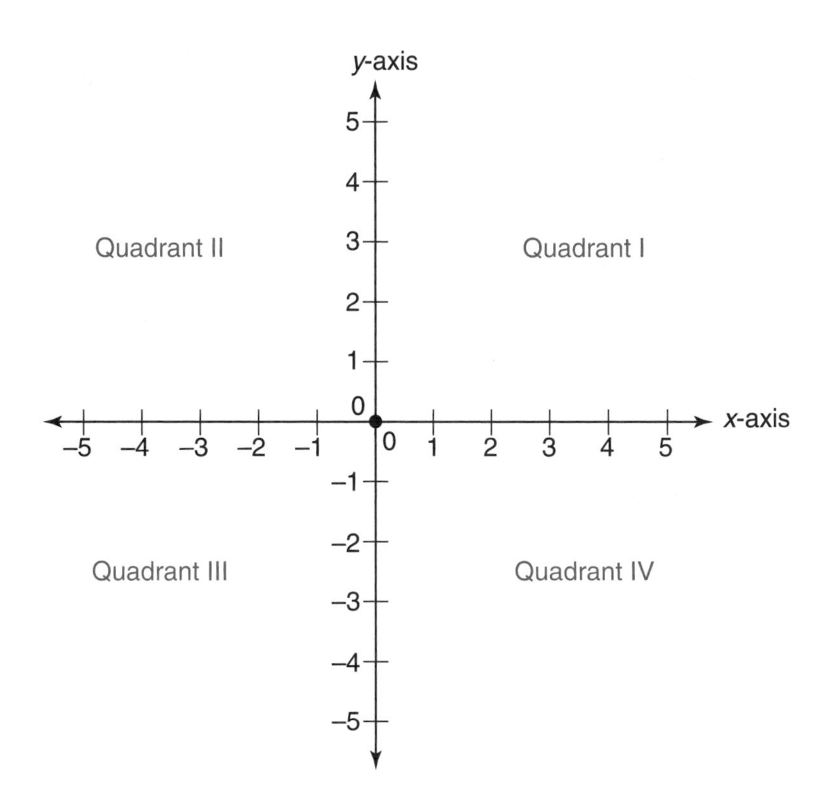

You need to know these rules to graph:

- If you move to the right of zero, the numbers are positive.
- If you move to the left of zero, the numbers are negative.
- If you move up from zero, the numbers are positive.
- If you move down from zero, the numbers are negative.

When you graph points, you use positive numbers, negative numbers, and zero. The fancy name for these numbers is **integers**. Graphing points and moving in the positive or negative direction will help us better visual the integers, which are positive and negative whole numbers!

To locate any point, use an ordered pair: $(x,y) \rightarrow$ (left or right, up or down).

To graph a point, always start at (0,0). Then remember x always comes first, and y comes second. So move left or right, and then up and down.

$\frac{x,y}{3,2} \rightarrow$ Move right three, and move up 2.

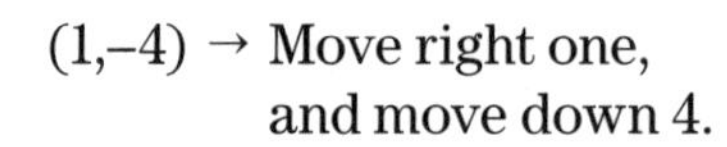
(1,–4) → Move right one, and move down 4.

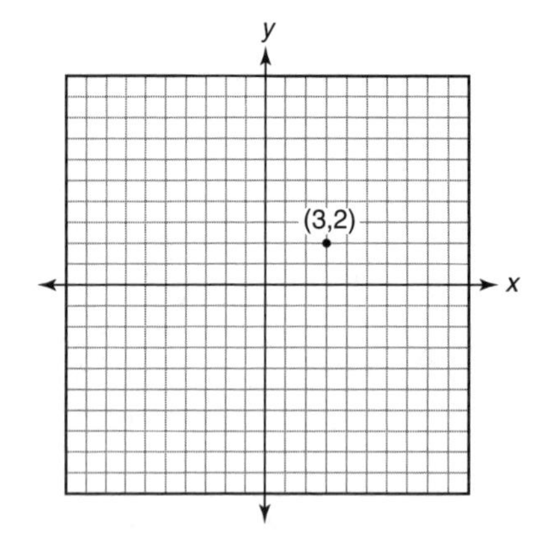

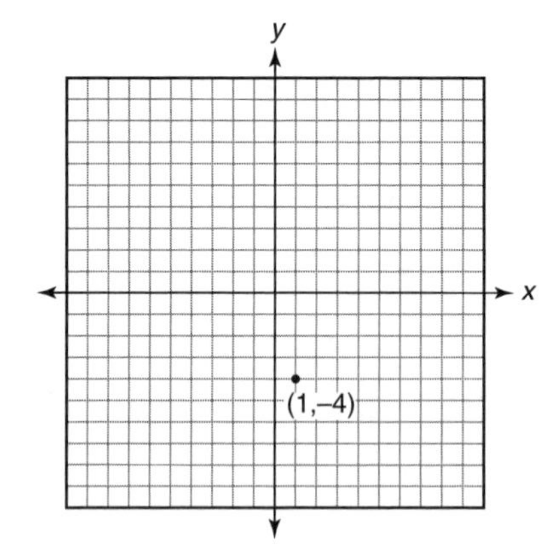

(–3,–4) → Move left 3, and move down 4.

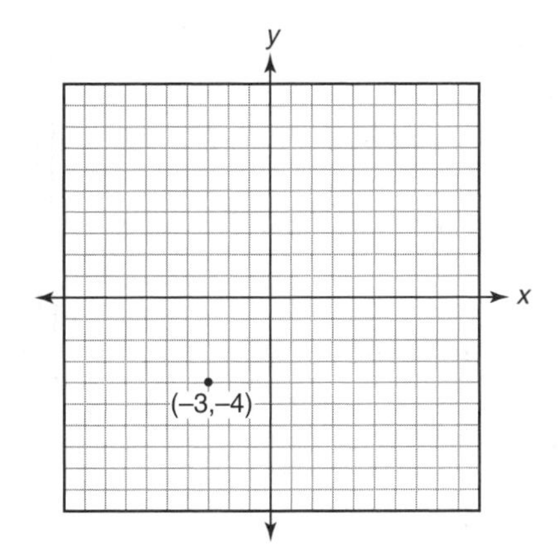

(0,5) → Do not move left or right, and then move up 5.

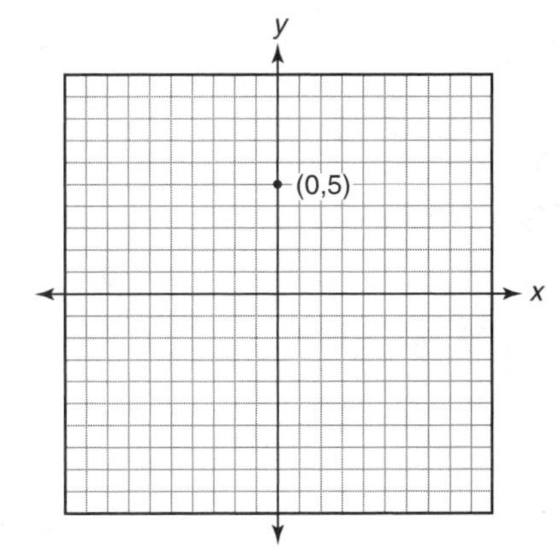

How would you get to your friend's house if your friend said, "Walk right 2 blocks and up 3 blocks"? From your house, move right 2 and up 3. That is the point (2,3)! See, it's painless!

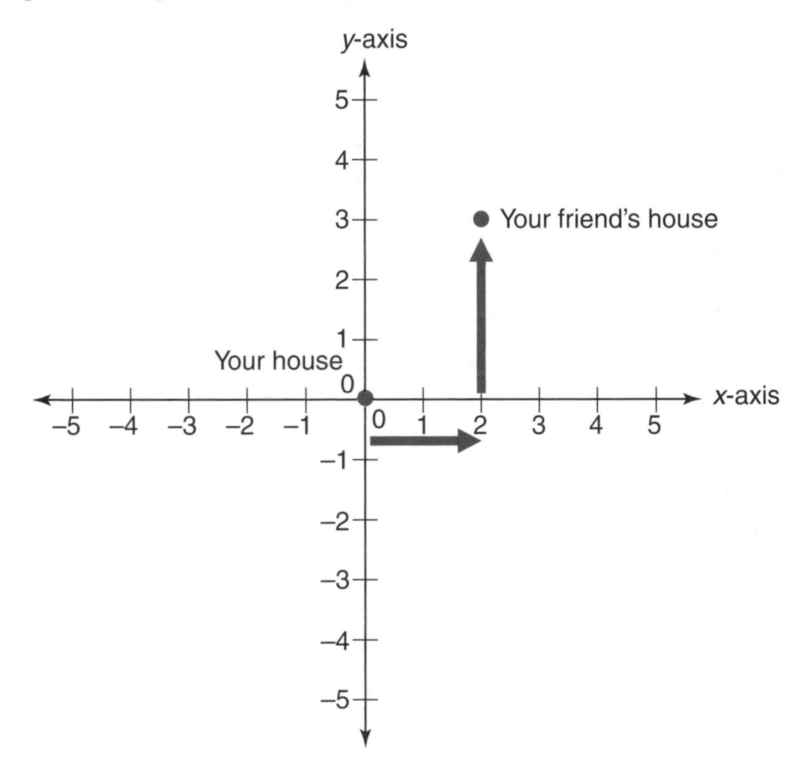

Caution—Major Mistake Territory!

Always move left or right and then move up or down.

Pay special attention to your graphing when points fall on the axes.

(0,5) → Don't move left or right. Then move up 5.

(3,0) → Move right 3. Do not move up or down.

(–3,0) → Move left 3. Do not move up or down.

(0,–5) → Do not move left or right. Then move down 5.

BRAIN TICKLERS
Set # 26

Write the coordinates of the following points.

1. *A*______
2. *B*______
3. *C*______
4. *D*______
5. *E*______
6. *F*______

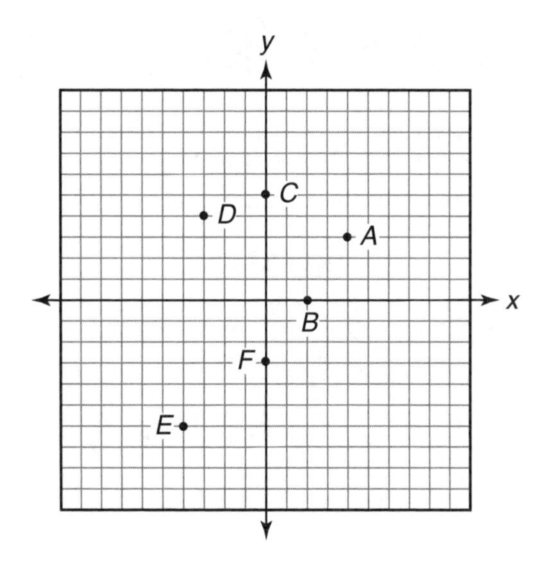

On graph paper, plot and label the following points. Also, name which quadrant each point is in.

7. $A(4,2)$ 8. $B(-3,5)$ 9. $C(-2,-4)$ 10. $D(4,-1)$
11. $E(5,0)$

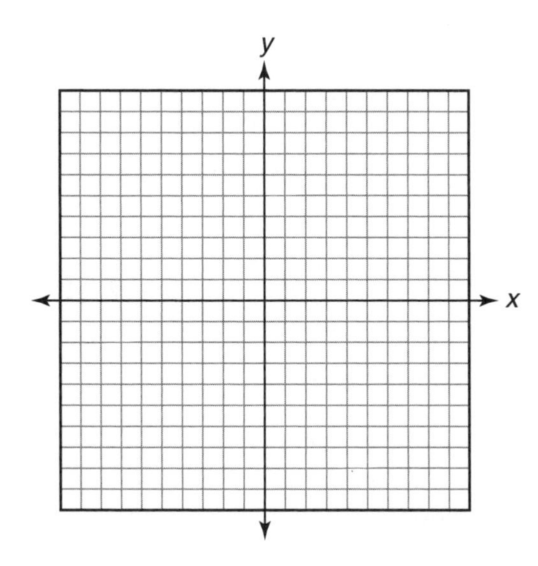

(Answers are on page 118.)

WHAT ARE INTEGERS?

Positive and negative numbers are everywhere. The temperature went up 5 degrees. You lost 10 pounds. Your teacher gave you 2 bonus points. The football team had a loss of 10 yards.

INTEGERS

Positive numbers	1, 2, 3, 4, 5, 6, 7, 8, . . .
Zero	0
Negative numbers	–1, –2, –3, –4, –5, –6, –7, . . .

The set of numbers shown in the table is called the **integers**. So what are the integers?

DEFINITION

Integers are the set of positive and negative whole numbers. Remember that the whole numbers include zero!

EXAMPLES OF INTEGERS

. . . –4, –3, –2, –1, 0, 1, 2, 3, 4, . . .

NOT INTEGERS

$$\left.\begin{array}{l} -3.5 \\ -\frac{1}{2} \\ 13.25 \\ \frac{3}{4} \\ 0.123456\ldots \end{array}\right\} \text{None of these are whole numbers!!!}$$

EXAMPLES IN YOUR LIFE

You deposit $50 in the bank (+50).

You withdraw $25 from the ATM (–25).

In golf, you hit 4 under par (–4).

The temperature rose 15°F (+15).

You owe your mom $10 (–10).

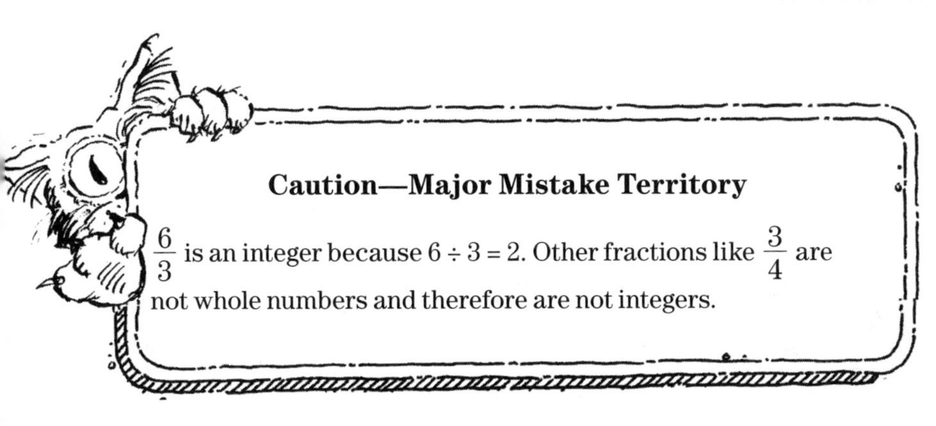

Caution—Major Mistake Territory

$\frac{6}{3}$ is an integer because $6 \div 3 = 2$. Other fractions like $\frac{3}{4}$ are not whole numbers and therefore are not integers.

Here are some situations in your life that represent integers. You probably didn't even realize they are integers!

- A loss of 5 yards → –5
- A pay cut of $6,000 → –6000
- 9 degrees above zero → 9
- A $500 bonus → 500
- 10 degrees below 0 → –10

Just remember that the name integers is just describing a certain type of number. Any positive or negative whole number is an integer.

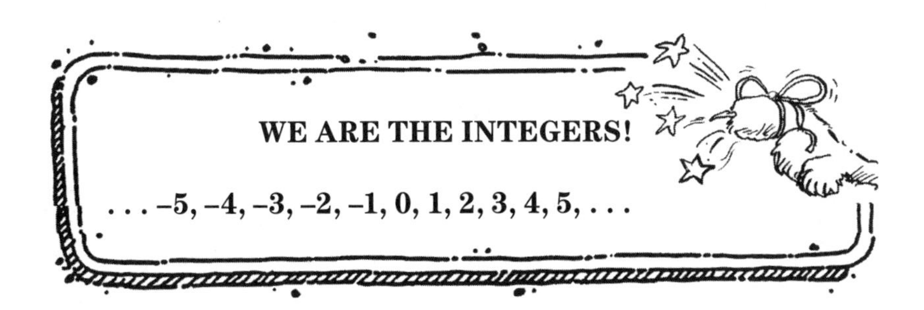

WE ARE THE INTEGERS!

. . . –5, –4, –3, –2, –1, 0, 1, 2, 3, 4, 5, . . .

BRAIN TICKLERS
Set # 27

1. Circle all of the integers.
   $-3.5 \quad 0 \quad 4 \quad -5 \quad -4.2 \quad 1\frac{2}{5} \quad -12 \quad 28$
2. Explain why 3.5 is not an integer.

Write an integer to represent each situation.

3. The stock market dropped 23 points.
4. Your apartment is on the 13th floor.
5. The temperature is 4 degrees below zero.
6. You withdrew $200.

(Answers are on page 118.)

COMPARING AND ORDERING INTEGERS

You are 13 years old. Your sister is 16 years old. Who is older? Your sister, of course! When you answered that question, you ordered integers.

What are integers? They are positive and negative whole numbers. Let's look at the number line to see how we can put them in order.

The number line below shows the integers from –5 to 5. Remember from graphing that positive numbers are to the right of zero and negative numbers are to the left of zero.

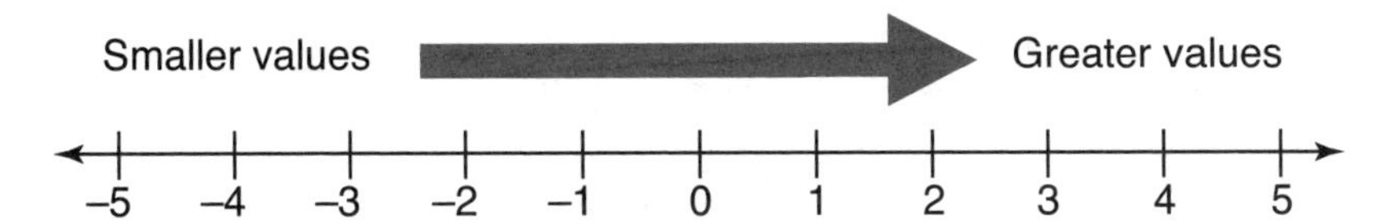

When you are reading a number line, the numbers on the number line increase in value as you move from left to right. Therefore, –2 is larger than –5, 2 is greater than –1, and 5 is greater than 3.

Let's look at a list of low temperatures recorded during the last 6 months in Rochester, NY, from September to February.

SEPT.	OCT.	NOV.	DEC.	JAN.	FEB.
52°	48°	29°	–2°	11°	–12°

To order these temperatures from least to greatest, we have to think of how they would fall on the number line. From least to greatest, the numbers are –12, –2, 11, 29, 48, 52.

In math, when comparing two numbers, we use inequalities symbols to indicate which value is greater than or less than another value. The symbols are:

$>$ Greater than $<$ Less than

- $-5 < -3$ is read as "negative 5 is less than negative 3."
- $8 > -5$ is read as "eight is greater than –5."

Caution—Major Mistake Territory!

When comparing two negative numbers, it is easy to get confused about which is larger or smaller. When dealing with negative numbers, think of the number line. The closer to 0 the number is, the larger it is.

So –5 is smaller than –3, and
–4 is smaller than –1.

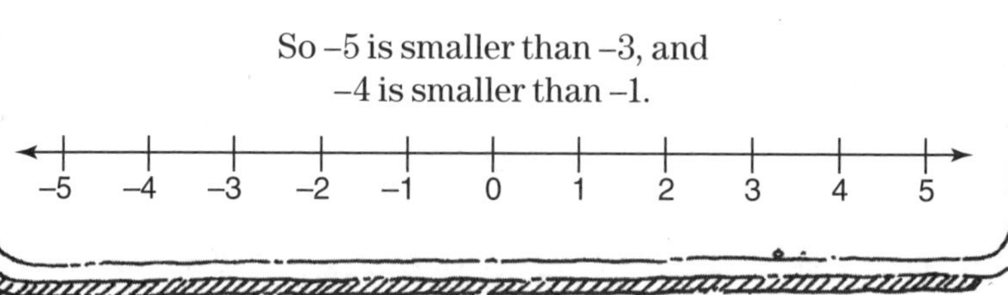

Putting numbers in order and comparing them is painless! Just remember, would you rather have $20 in your pocket or be in debt $5? Which temperature is warmer, –3° or 16°? If you think of a situation in your life that deals with positive and negative numbers, putting them in order will be easier.

BRAIN TICKLERS
Set # 28

1. Order from least to greatest.
   –5, 10, 23, –15, –24, 17, 32

2. Order from greatest to least.
   –24, –15, 0, 16, –42, 38, 50

3. Which integer is between –4 and –6?

Compare using <, >, or =.

4. –8____–12   5. 15____–6   6. –2____0

(Answers are on page 118.)

ABSOLUTE VALUE AND OPPOSITES

What is the opposite of black? White! What is the opposite of up? Down! You already know opposites. Well, you can also take the opposite of a number.

You know that the integers are . . . –3, –2, –1, 0, 1, 2, 3 . . . positive and negative whole numbers. We can also say that the integers are the set of whole numbers and their opposites. What do you think the opposite of 3 is? If you said –3, you are right! Let's see why.

Opposites, also known as **additive inverses**, are numbers that are the same distance from 0 but on different sides of the number line.

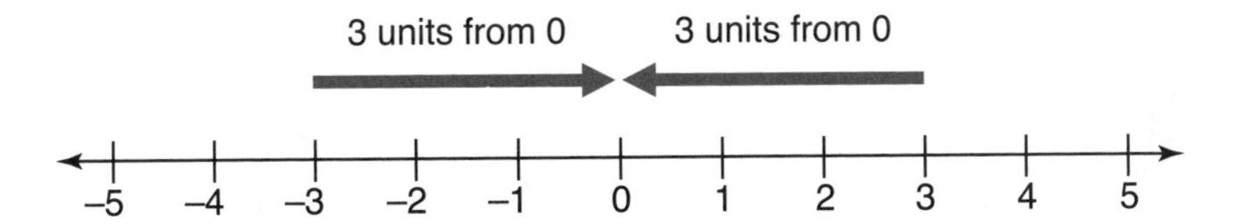

The number –3 is three units from zero, and the number 3 is also three units from zero. Therefore 3 and –3 are opposites or additive inverses of each other.

- What is the opposite of 6? → –6
- Find the additive inverse of –15. → 15

Finding the additive inverse is easy. Just remember it is the opposite of the number!

A number's **absolute value** is its distance from zero on the number line.

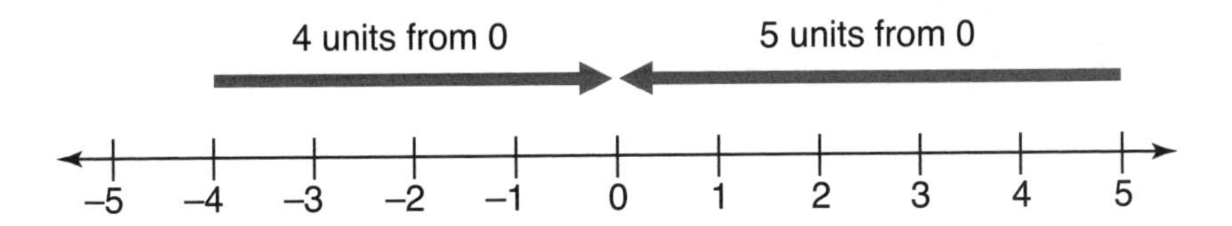

-4 is four units from zero, so the absolute value of -4 is 4. We write absolute value using bar notation.

This notation represent the absolute value of -4.

$|-4| \quad = 4$

5 is five units from zero, so the absolute value of 5 is 5. You would write this as $|5|$.

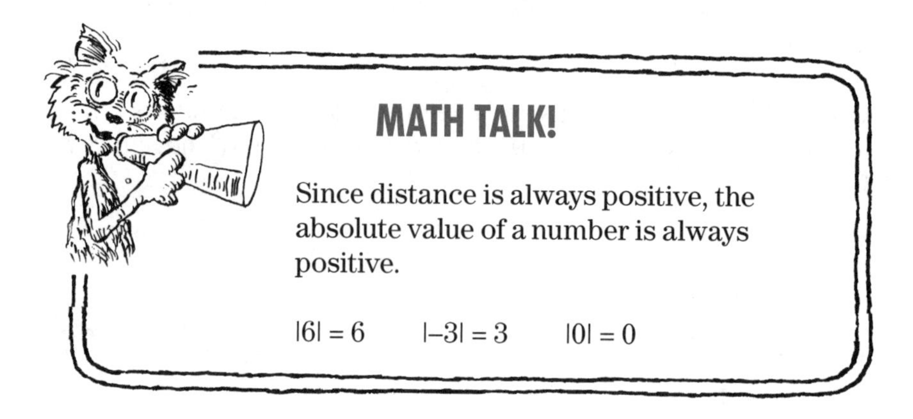

MATH TALK!

Since distance is always positive, the absolute value of a number is always positive.

$|6| = 6 \qquad |-3| = 3 \qquad |0| = 0$

If you remember these rules, solving problems involving absolute value is painless! Try these.

Example 1

$\|9\| + \|7\|$	Find the absolute value of each number.
$9 + 7$	Now add.
16	

Example 2

$\|20 - 5\|$	Since there is math to do inside the absolute value bars, do it!
$\|15\|$	Now take the absolute value of 15.
15	

Caution—Major Mistake Territory!

You can take the absolute value of only a single number! If an operation is inside the absolute value bars, simplify first and then take the absolute value of the number!

Example 3

$|-7 - 5|$ Simplify first!
$|-12|$ Take the absolute value of -12.
12

BRAIN TICKLERS
Set # 29

Find the additive inverse of each number.

1. -10 2. 14 3. -26 4. 0

Evaluate.

5. $|-16|$ 6. $|5| + |-10|$ 7. $|22 - 9|$
8. $|10 - 10|$

Compare using <, >, or =.

9. $|-8|$ ____ $|7|$ 10. $|-14|$ ____ $-|15|$ 11. $|3 + 2|$ ____ $|-5|$

(Answers are on page 118.)

ADDING INTEGERS

Adding integers can be easy as pie. Remember two rules and you'll be fine!

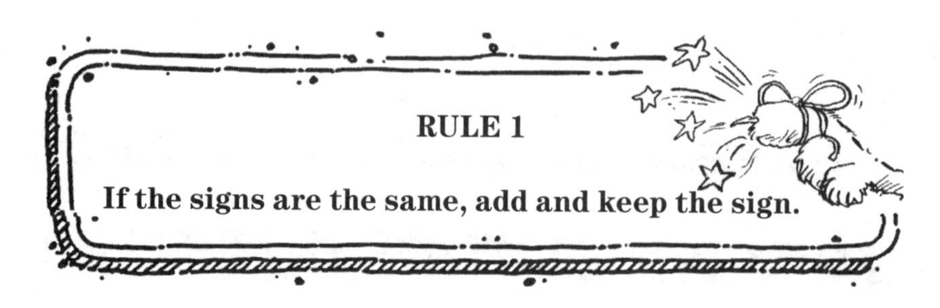

8 + 7 = 15 14 + 10 = 24	Both numbers are positive. Add. The answer is also positive.
–2 + –3 = –5 –4 + –8 = –12	Both numbers are negative. Add. The answer is also negative.

RULE 2

If the signs are different, subtract and keep the sign of the number with the greatest absolute value.

8 + (–2) = 6	The signs are different, so subtract. There are more positives (8), so the answer is positive.
–12 + 7 = –5	The signs are different, so subtract. There are more negatives (–12), so the answer is negative.
5 + (–9) = –4	The signs are different, so subtract. There are more negatives (–9), so the answer is negative.
7 + (–4) = 3	The signs are different, so subtract. There are more positives (7), so the answer is positive.

MATH TALK!

Keep the sign of the number with the greatest absolute value. This really means, do you have more negatives or positives? Keep that sign!

When adding integers, it is helpful to think of a story.

6 + (–2) You have \$6, and you bought a pack of gum for \$2. How much money do you have left? \$4

–12 + 5 You owe your sister \$12. You pay her back \$5. Now you owe her \$7! (–7)

–14 + –15 The football team lost 14 yards on the first play and 15 yards on the second play. Together they lost 29 yards! (–29)

Make up a story—it can be silly. This will make adding integers super, super easy!

BRAIN TICKLERS
Set # 30

Add:

1. 16 + 3
2. 4 + (–6)
3. –5 + 8
4. –14 + –20
5. –33 + 16
6. –14 + 18
7. 9 + (–9)
8. –49 + –7
9. 36 + 18
10. 38 + –14

Use the given numbers to make each equation true. Numbers may be used more than once.

4 7 8 –4 –7 –8

11. ____ + ____ = –3

12. ____ + ____ = –1

13. ____ + ____ = –11

14. ____ + ____ = 3

15. ____ + ____ = –4

16. ____ + ____ = 4

17. ____ + ____ = 1

18. ____ + ____ = –15

(Answers are on page 119.)

SUBTRACTING INTEGERS

Now that you are a whiz at adding integers, subtracting integers will be a breeze!

Subtracting a smaller number from a larger number is the same as finding how far apart the two numbers are on the number line. How can we subtract painlessly? Subtracting integers is the same as adding the opposite!

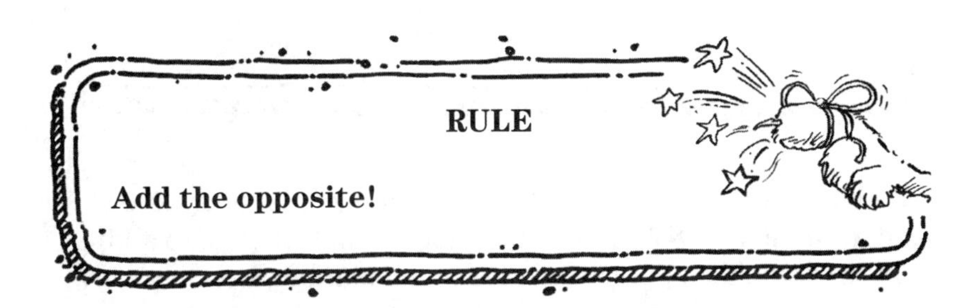

Exciting Examples

$4 - 7$	This is a subtraction problem, so add the opposite.
$4 + (-7)$	Change the subtraction to addition. The opposite of 7 is –7.
-3	The signs are different, so subtract. There are more negatives (–7), so the answer is negative.
$5 - (-8)$	This is a subtraction problem, so add the opposite.
$5 + 8$	Change the subtraction to addition. The opposite of –8 is 8.
13	Just add.
$-12 - (-6)$	This is a subtraction problem, so add the opposite.
$-12 + 6$	Change the subtraction to addition. The opposite of –6 is 6.
-6	The signs are different, so subtract. There are more negatives (–12), so the answer is negative.
$12 - (-5)$	This is a subtraction problem, so add the opposite.
$12 + 5$	Change the subtraction to addition. The opposite of –5 is 5.
17	Just add.

Caution—Major Mistake Territory!

Make sure you keep (K) the sign on the first number. Change the subtraction sign to addition (A). Then take the opposite (O) of the second number only!

$10 - (-5) = 10 + 5$
K A O

Another way to remember subtracting integers is to say **keep, add, opposite** or **KAO**. Keep the first number, change to addition, and take the opposite of the second number. Let's try a few more!

- $-12 - 8 \rightarrow -12 + (-8)$
  KAO
- $-15 - (-6) \rightarrow -15 + 6$
  KA O
- $25 - (-8) \rightarrow 25 + 8$
  K A O
- $4 - 8 \rightarrow 4 + (-8)$
  KAO

BRAIN TICKLERS
Set # 31

Perform the given operation.

1. $-6 - 5$

2. $-4 - (-10)$

3. $15 - 9$

4. $-12 - (-10)$

5. $-11 - 18$

6. $-26 - (-5)$

7. $17 - 25$

8. $-13 - 8 - (-6)$

9. $25 - 36 - 10$

10. Reed had a savings account with $125 in it. He withdrew $10 on Monday, $20 on Tuesday, and $15 on Wednesday. On Friday, he made a deposit of $35. How much money is in Reed's account after Friday's deposit?

(Answers are on page 119.)

MULTIPLYING AND DIVIDING INTEGERS

Do you like pizza? I know I like pizza! Did you know that pizza can help you multiply and divide?

Use this pizza rule to help you remember your rules for multiplying and dividing integers.

+	+	+	I like pizza (+). You like pizza (+). We agree (+)!
+	–	–	I like pizza (+). You don't like pizza (–). We disagree (–)!
–	+	–	I don't like pizza (–). You like pizza (+). We disagree (–)!
–	–	+	I don't like pizza (–). You don't like pizza (–). We agree (+)!

MATH TALK!

Signs same, answer positive
Signs different, answer negative

+	+	+	Signs same, answer positive
+	–	–	Signs different, answer negative
–	+	–	Signs different, answer negative
–	–	+	Signs same, answer positive

Pizza Rule Examples

$4(5) = 20$ I like pizza. You like pizza. We agree (answer positive).

$-20(5) = -100$ I don't like pizza. You like pizza. We disagree (answer negative).

$-6(-7) = 42$	I don't like pizza. You don't like pizza. We agree (answer positive).
$10(-6) = -60$	I like pizza. You don't like pizza. We disagree (answer negative).

Math Talk Examples

$\frac{-20}{5} = -4$	Signs different, answer negative
$\frac{-40}{-8} = 5$	Signs same, answer positive
$\frac{120}{10} = 10$	Signs same, answer positive
$\frac{60}{-6} = -10$	Signs different, answer negative

Caution—Major Mistake Territory!

Remember, any number multiplied by zero equals zero. So if a number is multiplied by zero, the integer rules do not apply!

$5(0) = 0$ $-4(0) = 0$ $-3(-4)(0) = 0$

The multiplication rules also apply to problems that involve more than two numbers. Just remember to use the rules for two numbers at a time.

Exciting Examples

$2(-3)(4)$	First multiply 2 times –3.
$-6(4)$	Then multiply –6 by 4.
-24	

$(-2)^3$	Since (–2) is cubed, write (–2) times itself three times.
$(-2)(-2)(-2)$	First multiply (–2)(–2).
$4(-2)$	Then multiply 4 times (–2).
-8	

BRAIN TICKLERS
Set # 32

1. $-12(3)$
2. $(15)(-8)$
3. $-76 \div (-4)$
4. $(-4)(-12)$
5. $150 \div -15$
6. $-3(2)(-2)$
7. $(-3)^3$
8. $\frac{-88}{11}$

Perform each of the indicated operations. To answer the puzzle, fill in the letter that corresponds to each answer.

T	T	O	S	R
$-12(3)$	$\frac{60}{-6}$	$-22(-5)$	$\frac{35}{7}$	$(-4)^2$

A	H	M	E	A
$\frac{-63}{-7}$	$-2(3)(-4)$	$\frac{-72}{9}$	$-5(0)$	$14(3)$

What song did the large pizza sing to the small pizza?

____ ____ ____ ____ ' ____   ____ ____ ____ ____ ____
–10   24   9   –36   5   42   –8   110   16   0

(Answers are on page 119.)

ORDER OF OPERATIONS WITH INTEGERS

Let's review the order of operations one more time!

- P	Simplify inside the parentheses.
- E	Evaluate the exponent.
- MD	Multiply and divide from left to right.
- AS	Add and subtract from left to right.

The neat thing about the order of operations is that it works for all numbers, both positive and negative! Since you are a pro with basic order of operations from Chapter Two, let's make simplifying problems with negatives painless!

Exciting Examples

$-12 - 9 - (-6 + 9)$	Simplify inside the parentheses.
$-12 - 9 - 3$	Subtract left to right; rewrite subtraction as "add the opposite."
$-12 + (-9) + (-3)$	Add $-12 + (-9)$.
$-21 + (-3)$	Add.
-24	

$-24 \div 6(2) - (-4)$	Since there is multiplication and division in the problem, work left to right, $-24 \div 6$.
$-4\,(2) - (-4)$	Next, multiply $-4(2)$. Remember the pizza rule!
$-8 - (-4)$	Since this is subtraction, add the opposite.
$-8 + 4$	Simplify.
-4	

$-3(-5) - 6(-2)$	Multiply left to right. Start with -3 times -5.
$15 - 6(-2)$	Next, multiply 6 times -2.
$15 - (-12)$	Since this is subtraction, add the opposite!
$15 + 12$	Add.
27	

$4(-2)^3$	Be careful! Remember to do exponents before you multiply! Rewrite $(-2)^3$ as a multiplication problem.
$4(-2)(-2)(-2)$	Multiply left to right, $4(-2)$.
$-8(-2)(-2)$	Multiply 8 times -2.
$16(-2)$	Simplify. Remember the rule. "I like pizza. You
-32	don't like pizza. We disagree!"

Caution—Major Mistake Territory!

Pay special attention to problems involving exponents and multiplication. $2(-3)^2$ means $(-3)^2$ should be simplified first because of the exponent.

So $2(-3)^2 = 2(9) = 18$.

BRAIN TICKLERS
Set # 33

Order it Up! Use the order of operations to simplify each of the following.

1. $-16 + (20 \times 6) \div (-6 + 2) - 8$
2. $-4(3) - 4(13)$
3. $-3(-2 + -5) + 8 + (-2)^2$
4. $-4(-3)^3 + 1$
5. $2(8 - 9) - (15 - 25)$
6. $-2 - 3 - 5 - 7 + 8 \div (-2)$

(Answers are on page 119.)

ONE-STEP EQUATIONS WITH INTEGERS

Solving one-step equations with positive and negative numbers is painless! You will either have to add or subtract and will have to multiply or divide. Remember these three steps, and you are all set!

Step 1: Undo add or subtract.

Step 2: Undo multiply or divide.

Step 3: Simplify.

The painless way to remember these rules is to use the opposite operation to get to the opposite side!

Let's just figure out what sign to use.

$x + (-6) = 9$ Use the painless way. Since the number is –6, add 6 to get to the opposite side!

$x - 7 = -14$ Since the problem says subtract 7, add 7 to both sides.

Let's look at a few addition and subtraction problems to see how easy solving equations can be!

Exciting Example

$x + 6 = -14$

Remember the goal is to get the variable (letter) by itself. So how do you get x by itself? Since it says +6, we have to subtract 6 from both sides.

$$\begin{array}{rcr} x - 6 & = & -14 \\ +6 & & +6 \\ \hline x & = & -8 \end{array}$$

Use your integer rules to add –14 + 6. See, it's painless!

Another Example

$x + 5 = -12$

$$\begin{array}{rl} x + 5 = -12 & \\ -5 = \ \ -5 & \\ \hline x \qquad = -17 & \end{array}$$

Since this problem says +5, you have to subtract 5 to get to the other side

Now use your integer rules → $-12 - 5 = -12 + -5$.

Brain Buster!

$x - (-8) = 12$

$$\begin{array}{l} x - (-8) = 12 \\ x + 8 = 12 \\ \ \ -8 \ \ -8 \\ \hline x \qquad = \ 4 \end{array}$$

Think of an easy way to rewrite this → $x + 8 = 12$.

Now it's painless! Since it says +8, subtract 8 from both sides.

Simplify.

MATH TALK!

Use the opposite sign to get to the opposite side of the equation.

$x + 10 = 8$ → Since +10, add −10 to get to the other side.

$x - 12 = -8$ → Since −12, add 12 to get to the other side!

Now let's try multiplication and division.

Example 1

$$\frac{-2x}{-2} = \frac{24}{-2}$$

The equation reads "negative two times x equals 24." Since the equation says to multiply, undo by dividing! Now simplify; 24 divided by $-2 = -12$! You go math whiz!

Example 2

$\frac{x}{4} = -3$ The equation reads "x divided by 4 equals –3." How do you undo division? Multiply.
To undo division by 4, multiply by 4.

$$4 \cdot \frac{x}{4} = -3(4)$$

So $x = -12$.

To make solving equations painless, just remember to use the opposite operation!

- To undo addition, subtract.
- To undo subtraction, add.
- If it is a multiplication problem, undo by dividing.
- If it is a division problem, undo by multiplying.

BRAIN TICKLERS
Set # 34

Solve each of the following.

1. $x + 10 = 5$
2. $n - 12 = -8$
3. $x + 8 = -13$
4. $p - 12 = -22$
5. $-6 + m = -14$
6. $-8 + t = 22$
7. $2x = -16$
8. $-4x = -36$
9. $-5x = 100$
10. $\frac{x}{4} = -7$
11. $\frac{x}{-2} = -6$
12. $\frac{x}{-5} = 12$

(Answers are on page 120.)

BRAIN TICKLERS—THE ANSWERS

Set # 26, page 94

1. $A(4, 3)$
2. $B(2, 0)$
3. $C(0, 5)$
4. $D(-3, 4)$
5. $E(-4, -6)$
6. $F(0, -3)$

Set # 27, page 98

1. 0, 4, –5, –12, 28
2. 3.5 is not a whole number.
3. –23
4. 13
5. –4
6. –200

Set # 28, page 100

1. –24, –15, –5, 10, 17, 23, 32
2. 50, 38, 16, 0, –15, –24, –42
3. –5
4. $>$
5. $>$
6. $<$

Set # 29, page 103

1. 10
2. –14
3. 26
4. 0
5. 16
6. 15
7. 13
8. 0
9. $>$
10. $>$
11. $=$

Set # 30, page 105

1. 19
2. −2
3. 3
4. −34
5. −17
6. 4
7. 0
8. −56
9. 54
10. 24
11. −7 + 4
12. −8 + 7
13. −7 + (−4)
14. 7 + (−4)
15. −8 + 4
16. 8 + (−4)
17. 8 + (−7)
18. −8 + (−7)

Set # 31, page 109

1. −11
2. 6
3. 6
4. −2
5. −29
6. −21
7. −8
8. −15
9. −21
10. $115

Set # 32, page 112

1. −36
2. −120
3. 19
4. 48
5. −10
6. 12
7. −27
8. −8

Solution to puzzle: That's Amore

Set # 33, page 114

1. −54
2. −64
3. 33
4. 109
5. 8
6. −21

Set # 34, page 117

1. $x = -5$
2. $n = 4$
3. $x = -21$
4. $p = -10$
5. $m = -8$
6. $t = 30$
7. $x = -8$
8. $x = 9$
9. $x = -20$
10. $x = -28$
11. $x = 12$
12. $x = -60$

CHAPTER SIX

Exponents and Roots

EXPONENTS

Exponents are a shorthand way of writing repeated multiplication. For example $2 \cdot 2 \cdot 2 \cdot 2 \cdot 2$ is read as 2 times 2 times 2 times 2 times 2. This multiplication problem can also be written in **exponential form**.

$2 \cdot 2 \cdot 2 \cdot 2 \cdot 2 = 2^5 \rightarrow 2^5$ means the number *2 is a* factor *5 times*

If a number is written in exponential form, the exponent tells how many times the base is used as a factor.

2 is the base → 2^5 ← 5 is the exponent

MATH TALK!

Here are a few ways to say and write exponential expressions

3^2	4^3	a^4
3 times 3 Three to the second power 3 squared	4 times 4 times 4 Four to the third power Four cubed	*a* times *a* times *a* times *a* *a* to the fourth power

To evaluate exponents, just remember that using exponents is just a short way of writing repeated multiplication.

$3^2 = 3 \cdot 3 = 9$ $\quad$ $3^3 = 3 \cdot 3 \cdot 3 = 27$ $\quad$ $3^4 = 3 \cdot 3 \cdot 3 \cdot 3 = 81$

Caution—Major Mistake Territory!

5^2 means 5 times 5, which equals 25.

5^2 does not mean 5 times 2, which equals ten.

The exponent tells you how many times to multiply the base by itself.

We can also use exponents with negative numbers.

$(-2)^2 = (-2)(-2) = 4$ $(-3)^2 = (-3)(-3) = 9$ $(-10)^2 = (-10)(-10) = 100$

When you square a negative number, the answer is always positive.

$(-2)^3 = (-2)(-2)(-2) = -8$ $(-3)^3 = (-3)(-3)(-3) = -27$
$(-4)^3 = (-4)(-4)(-4) = -64$

When you raise a negative number to the third power, or cube it, the answer is always negative.

RULES FOR EXPONENTS OF NEGATIVE NUMBERS

A negative number raised to an even power is always a positive number.
A negative number raised to an odd power is always a negative number.

$(-2)^2 = (-2)(-2) = 4$
$(-2)^3 = (-2)(-2)(-2) = -8$
$(-2)^4 = (-2)(-2)(-2)(-2) = 16$
$(-2)^5 = (-2)(-2)(-2)(-2)(-2) = -32$
$(-1)^{100} = 1$
$(-1)^{89} = -1$

Another important exponent rule is the **zero power rule**. Any number raised to the zero power equals 1.

$3^0 = 1$ $4^0 = 1$ $25^0 = 1$ $(-2)^0 = 1$

BRAIN TICKLERS
Set # 35

Evaluate each of the following.

1. 2^3 2. 4^2 3. 8^2 4. 10^3 5. 7^0
6. $(-3)^3$ 7. $(-5)^2$ 8. $(-20)^0$

(Answers are on page 147.)

PROPERTIES OF EXPONENTS

MULTIPLYING NUMBERS WITH THE SAME BASE

Exponents are a shorthand way of writing repeated multiplication. We can group the factors of the powers, such as 2^4, in different ways. Look at the following:

$2 \cdot 2 \cdot 2 \cdot 2 = 2^4$
$(2 \cdot 2) \cdot (2 \cdot 2) = 2^2 \cdot 2^2 = 2^4$
$2 \cdot (2 \cdot 2 \cdot 2) = 2^1 \cdot 2^3 = 2^4$

When you multiply powers with the same base, keep the base and add the exponents.

Exciting Examples

$52 \cdot 53$ → Bases are the same. Add the exponents!
5^{2+3}
5^5
$4^{10} \cdot 4^8$
4^{10+8}
4^{18}
$x^5 \cdot x^{12}$
x^{5+12}
x^{17}
$a^{15} \cdot a^{-3}$
$a^{15+(-3)}$
a^{17}

MATH TALK!

Remember what you know about exponents. Write out $6^4 \cdot 6^2$ in standard form.

$$6^4 \cdot 6^2$$
$$(6 \cdot 6 \cdot 6 \cdot 6) \cdot (6 \cdot 6)$$

What does $6^4 \cdot 6^2$ equal? 6^6

The shortcut—add the exponents

Let's see what happens when we divide powers with the same base.

$$\frac{4^7}{4^3} = \frac{4 \cdot 4 \cdot 4 \cdot 4 \cdot 4 \cdot 4 \cdot 4}{4 \cdot 4 \cdot 4} = \frac{\cancel{4} \cdot \cancel{4} \cdot \cancel{4} \cdot 4 \cdot 4 \cdot 4 \cdot 4}{\cancel{4} \cdot \cancel{4} \cdot \cancel{4}} = 4 \cdot 4 \cdot 4 \cdot 4 = 4^4$$

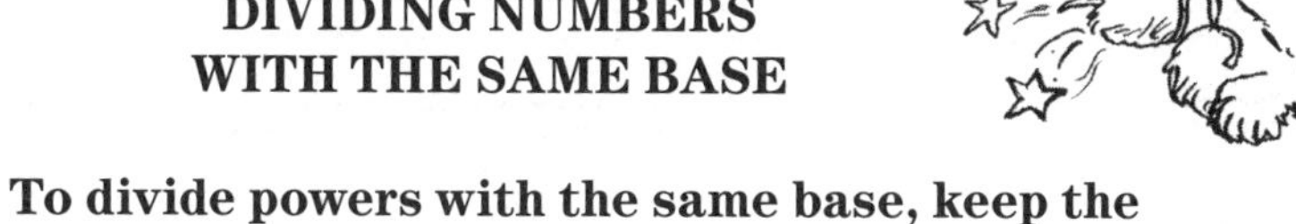

DIVIDING NUMBERS WITH THE SAME BASE

To divide powers with the same base, keep the base and subtract the exponents!

More Examples

$\frac{5^{12}}{5^4}$

5^{12-4}

5^8

$\frac{7^{20}}{7^{14}}$

7^{20-14}

7^6

$\frac{x^9}{x^4}$

x^{9-4}

x^5

Let's review!

- To multiply powers with the same base, keep the base and add the exponents.
- To divide powers with the same base, keep the base and subtract the exponents.

MATH TALK!

When multiplying exponents with the same base, add the exponents.
When dividing exponents with the same base, subtract the exponents.

Hints to help you:
Times sign × → rotated looks like +
Division sign ÷ → has minus (–) in it

To see what happens when you **raise a power to a power**, we are going to use what we know about writing out exponents and the order of operations.

Example 1

$(2^3)^4$ Inside the parentheses, what does 2^3 mean?

$(2 \cdot 2 \cdot 2)^4$ What does an exponent of 4 mean?

$(2 \cdot 2 \cdot 2) \cdot (2 \cdot 2 \cdot 2) \cdot (2 \cdot 2 \cdot 2) \cdot (2 \cdot 2 \cdot 2)$ Write this using one exponent.
$= 2^{12}$

Example 2

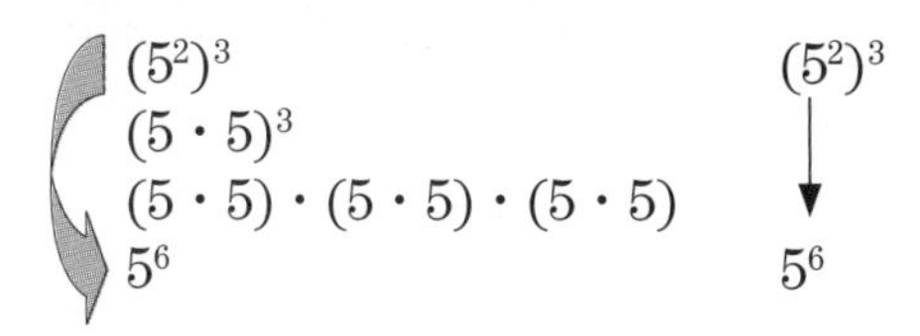

$(5^2)^3$
$(5 \cdot 5)^3$
$(5 \cdot 5) \cdot (5 \cdot 5) \cdot (5 \cdot 5)$
5^6

$(5^2)^3$
5^6

RAISING A POWER TO A POWER

To raise a power to a power, keep the base and multiply the exponents!

Practice

$(4^2)^4$	$(6^4)^5$	$(x^{12})^2$
$4^{2(4)}$	$6^{4(5)}$	$x^{12(2)}$
4^8	6^{20}	x^{24}

Caution—Major Mistake Territory!

You have to keep in mind the difference between multiplying exponents with the same base and raising a power to a power.

$2^3 \cdot 2^5$ = when you multiply, you add the exponents = 2^8

$(2^3)^5$ = power to a power, you multiply the exponents = 2^{15}

EXPONENT RULES

Property	Rule	Example
Multiply powers with the same base	Keep the base, add the exponents	$4^5 \cdot 4^6 = 4^{11}$
Divide powers with the same base	Keep the base, subtract the exponents	$\frac{7^5}{7^3} = 7^2$
Raise a power to a power	Keep the base, multiply the exponents	$(6^5)^3 = 6^{15}$

BRAIN TICKLERS
Set # 36

Multiply.

1. $5^6 \cdot 5^4$ 2. $7^{10} \cdot 7^5$ 3. $x^4 \cdot x^{11}$

Divide.

4. $\frac{5^9}{5^6}$ 5. $\frac{8^{15}}{8^4}$ 6. $\frac{a^{18}}{a^2}$

Simplify.

7. $(3^4)^3$ 8. $(4^5)^2$ 9. $(x^5)^3$

Multiply. Write the product as one power.

10. $10^5 \cdot 10^5$ 11. $m^4 \cdot m^{-2}$ 12. $15 \cdot 15^3$

Divide. Write the quotient as one power.

13. $\frac{11^8}{11^7}$ 14. $\frac{14^5}{14^2}$ 15. $\frac{b^{14}}{b^7}$

(Answers are on page 147.)

NEGATIVE EXPONENTS

We just looked at properties of exponents in the last chapter. Remember, an exponent of a number tells you how many times the base is multiplied by itself. So $4^3 = 4 \cdot 4 \cdot 4$. But what about something like

$$4^{-3}$$

What happens if the exponent is negative?

Let's look for a pattern in the table, using what we already know about exponents.

10^2	10^1	10^0	10^{-1}	10^{-2}	10^{-3}
$10 \cdot 10$	10	1 (remember, anything to the 0 power = 1)	$\frac{1}{10}$	$\frac{1}{10 \cdot 10}$	$\frac{1}{10 \cdot 10 \cdot 10}$
100	10	1	$\frac{1}{10} = 0.1$	$\frac{1}{100} = 0.01$	$\frac{1}{1000} = 0.001$

Notice: ÷ 10 ÷ 10 ÷ 10 ÷ 10 ÷ 10

When the exponent is positive, that tells us how many times to multiply the number by itself. So when the exponent is negative, that tells us how many times to divide by the number.

For example:

$$5^{-1} = 1 \div 5 = \frac{1}{5}$$

$$5^{-2} = 1 \div (5 \div 5) = \frac{1}{5^2}$$

$$5^{-3} = 1 \div (5 \div 5 \div 5) = \frac{1}{5^3}$$

Another way to think about it is:

$$a^{-n} = \frac{1}{a^n}$$

When any number, except 0, has a negative exponent, take the **reciprocal** of the number with its opposite exponent.

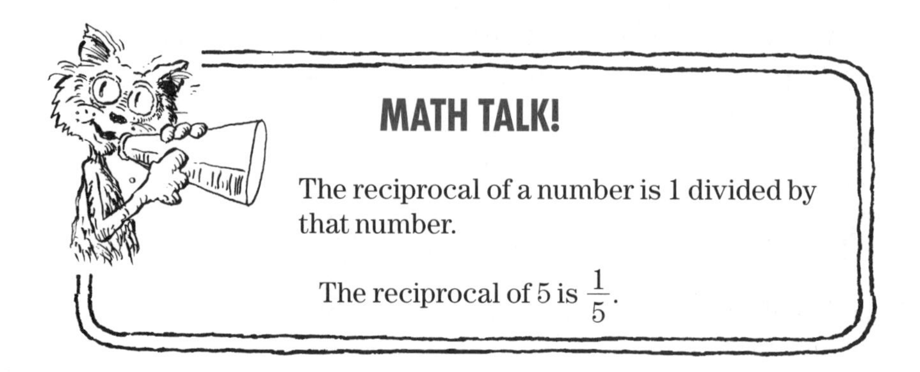

MATH TALK!

The reciprocal of a number is 1 divided by that number.

The reciprocal of 5 is $\frac{1}{5}$.

Let's look at a few more.

NEGATIVE EXPONENT	RECIPROCAL WITH POSITIVE EXPONENT	ANSWER
4^{-2}	$\frac{1}{4^2}$	$\frac{1}{4} \cdot \frac{1}{4} = \frac{1}{16}$
10^{-3}	$\frac{1}{10^3}$	$\frac{1}{10} \cdot \frac{1}{10} \cdot \frac{1}{10} = \frac{1}{1000}$
2^{-4}	$\frac{1}{2^4}$	$\frac{1}{2} \cdot \frac{1}{2} \cdot \frac{1}{2} \cdot \frac{1}{2} = \frac{1}{16}$
$(-2)^{-3}$	$\frac{1}{(-2)^3}$	$\frac{1}{-2} \cdot \frac{1}{-2} \cdot \frac{1}{-2} = \frac{1}{-8}$

Caution—Major Mistake Territory!

When dealing with negative exponents, remember to write the reciprocal of the number, and then change the sign of the *exponent*. If the original number is negative, that number stays negative when you take the reciprocal.

$$(-3)^{-1} = \frac{1}{(-3)^{1}}$$

$$(-3)^{-2} = \frac{1}{(-3)^{2}}$$

$$(-3)^{-3} = \frac{1}{(-3)^{3}}$$

Now you try! Evaluate:

1. 3^{-3} 2. 10^{-2} 3. $(-4)^{-2}$ 4. $(-2)^{-5}$

Answers:

1. $\frac{1}{3^3} = \frac{1}{27}$ 2. $\frac{1}{10^2} = \frac{1}{100}$ 3. $\frac{1}{(-4)^2} = \frac{1}{16}$ 4. $\frac{1}{(-2)^5} = \frac{1}{-32}$

BRAIN TICKLERS
Set # 37

Evaluate:

1. 10^{-5} 2. 3^{-4} 3. 2^{-6} 4. $(-3)^{-3}$
5. $(-7)^{-2}$

(Answers are on page 147.)

SCIENTIFIC NOTATION

Scientific notation is a shorthand way of expressing really large numbers or very small numbers. Understanding exponents makes understanding scientific notation easy! There are 2 parts to keep in mind when writing a number in scientific notation.

$$2.5 \times 10^5$$

The first number has to be greater than or equal to 1 and less than 10

The second number is 10 to a power. The power will show how many places to move the decimal point.

NUMBERS WRITTEN IN SCIENTIFIC NOTATION	NUMBERS NOT IN SCIENTIFIC NOTATION AND REASON		
3.2×10^3	0.1×10^3	→	First number is less than 1
4×10^{-4}	12×10^6	→	First number larger than 10
9.65×10^{23}	3×1^2	→	Second number has to be times 10 to a power

To translate a number **from scientific notation to standard form**, you need to look at the exponent on the 10. If the exponent is positive, move the decimal to the right. If the exponent is negative, move the decimal to the left. You move the decimal point as many times as the exponent indicates.

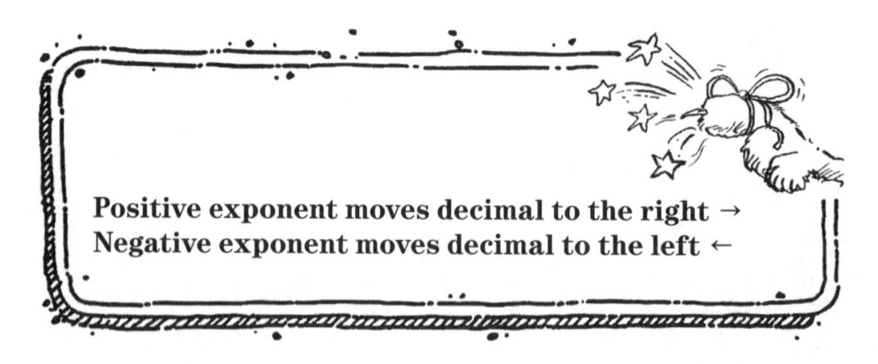

Positive exponent moves decimal to the right →
Negative exponent moves decimal to the left ←

Example 1

Write 3.25×10^5 in standard form.

10^5 is positive, so move the decimal 5 places to the right.

3.25×10^5
3.25000
→
325,000.

Example 2

Write 4.5×10^{-3} in standard form.

10^{-3} is negative, so move the decimal 3 places to the left.

4.5×10^{-3}
004.5
←
0.0045

Let's summarize.

- Write the digits with the decimal point placed after the first digit.
- Write × 10 to a power. The power shows how many places to move the decimal point.

SCIENTIFIC NOTATION	STANDARD FORM	SCIENTIFIC NOTATION	STANDARD FORM
2.3×10^2	230	4.5×10^{-1}	0.45
2.3×10^3	2,300	4.5×10^{-2}	0.045
2.3×10^4	23,000	4.5×10^{-3}	0.0045

Now let's go the other way. Let's translate a number **from standard form to scientific notation**.

Example 3

Take the number 25,000. Put the decimal after the first digit, keep all other different digits, but drop the repeated zeros.

2.5000

2.5 becomes the coefficient, which is the first part of the scientific notation form. → $2.5 \times 10^{?}$

Find the exponent by counting the number of places from the decimal to the end of the number. Since we will be moving right, the exponent will be positive.

In 25,000, there are 4 places. Therefore we write 25,000 as 2.5×10^{4}.

Write the number 0.0032 in scientific notation.

Example 4

Let's try a decimal. Take the number 0.0032. Put the decimal after the first digit.

0003.2

3.2 becomes the coefficient → $3.2 \times 10^{?}$. Find the exponent by counting the number of places from the new decimal to the decimal in the original number. Since we will be moving left, the exponent will be negative.

In 003.2, there are 3 places. Therefore we write 0.0032 as 3.2×10^{-3}.

LET'S REVIEW SCIENTIFIC NOTATION!

Place the decimal point after the **first digit**	Express 5245 in scientific notation 5245 → 5.245	Express 0.00045 in scientific notation 0.00045 → 4.5
Write the new number × 10	$5.245 \times 10^?$	$4.5 \times 10^?$
To find the power of 10, count the number of places from the decimal to get the original number: (large number → positive exponent) (decimal number → negative exponent)	5.245 → 5245 Decimal moved 3 places right	0004.5 → 0.00045 Decimal moved 4 places left
Scientific Notation	$\mathbf{5.245 \times 10^3}$	$\mathbf{4.5 \times 10^{-4}}$

BRAIN TICKLERS
Set # 38

Write each number in standard form.

1. 2.4×10^7
2. 3.65×10^{10}
3. 7.102×10^4
4. 5.2×10^{-3}
5. 2.9×10^{-1}
6. 4.56×10^{-6}

Write each number in scientific notation.

7. 234,000
8. 1,123
9. 307,000,000
10. 0.0005
11. 0.032
12. 0.00000456

(Answers are on page 147.)

SQUARES AND SQUARE ROOTS

When you square a number, you multiply the number by itself. So $1 \times 1 = 1$; $2 \times 2 = 4$; $3 \times 3 = 9$. The answers you get when you square a number are called perfect squares. Let's list our perfect squares up to 400 (20×20).

PERFECT SQUARES
(MEMORIZE THESE!)

1	(1 × 1)	121	(11 × 11)
4	(2 × 2)	144	(12 × 12)
9	(3 × 3)	169	(13 × 13)
16	(4 × 4)	196	(14 × 14)
25	(5 × 5)	225	(15 × 15)
36	(6 × 6)	256	(16 × 16)
49	(7 × 7)	289	(17 × 17)
64	(8 × 8)	324	(18 × 18)
81	(9 × 9)	361	(19 × 19)
100	(10 × 10)	400	(20 × 20)

When you take the **square root** of a number, you try to figure out what number, when multiplied by itself, will give this number. Think of it this way: you are "undoing" the perfect square.

For example, what is the square root of 36? Ask yourself, what number times itself is 36? It is 6! Here is how we write it:

$$\sqrt{36} = 6$$

This symbol $\sqrt{\ }$ is called a **radical sign**.

The number under the radical is called the **radicand**.

What is the square root of 100? Ask yourself, what number times itself equals 100? 10!

What is the square root of 49? What number, times itself, equals 49? 7!

You need to know the perfect squares in the chart above, because you can take the square root of these numbers. Let's make a new chart, showing all of your square roots.

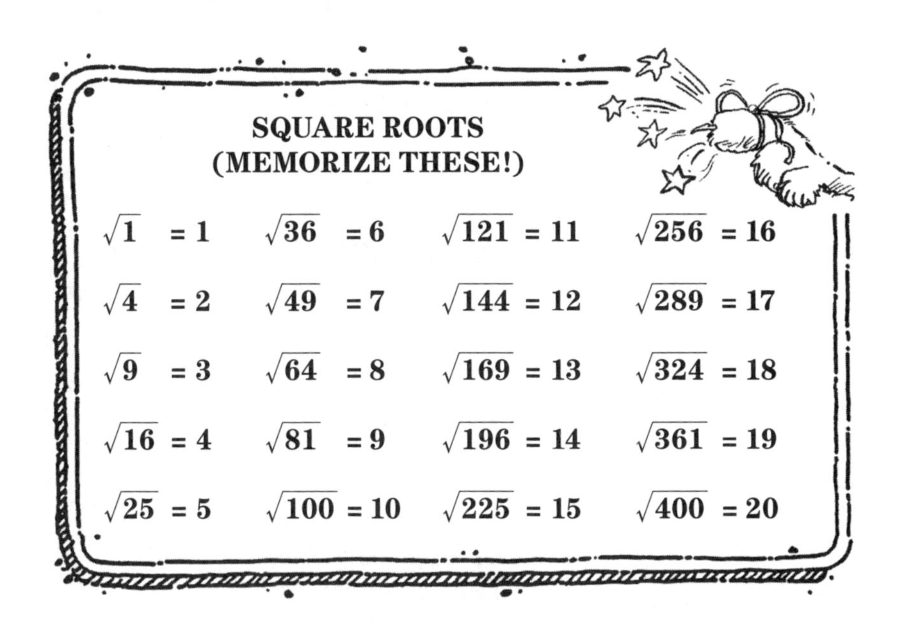

SQUARE ROOTS
(MEMORIZE THESE!)

$\sqrt{1} = 1$	$\sqrt{36} = 6$	$\sqrt{121} = 11$	$\sqrt{256} = 16$
$\sqrt{4} = 2$	$\sqrt{49} = 7$	$\sqrt{144} = 12$	$\sqrt{289} = 17$
$\sqrt{9} = 3$	$\sqrt{64} = 8$	$\sqrt{169} = 13$	$\sqrt{324} = 18$
$\sqrt{16} = 4$	$\sqrt{81} = 9$	$\sqrt{196} = 14$	$\sqrt{361} = 19$
$\sqrt{25} = 5$	$\sqrt{100} = 10$	$\sqrt{225} = 15$	$\sqrt{400} = 20$

To help you visualize squares and square roots, think about the relationship between the area of a square and the length of the sides of a square.

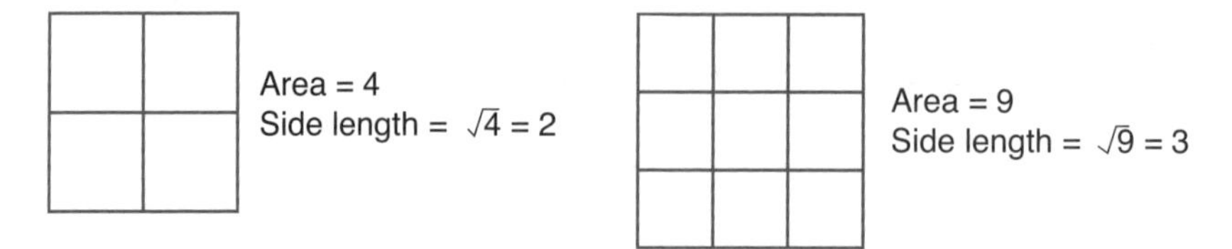

Every positive number has two square roots, one positive and one negative. The square root of 16 is 4 because 4(4) = 16. The other square root of 16 is –4 because –4(–4) also equals 16.

Technically, we can write $\sqrt{16} = \pm 4$, meaning "plus or minus" 4.

Usually, we give only the positive answer when finding the square root. Your calculator will give you only the positive answer. However, it is important to realize that there are two answers.

As the mathematicians say, the positive answer is the **principal square root.** When you say the square root of 81 is 9, you are finding the principal square root. Unless you are asked to find two square roots, in this book, we will find only the positive square roots.

MATH TALK!

To square a number means to multiply the number by itself.
Three squared = $3^2 = 9$

To find the square root of a number means what number times itself will give that number? The square root of 9 is 3 because 3 times 3 equals 9.

$$\sqrt{9} = 3$$

BRAIN TICKLERS
Set # 39

1. List the perfect squares up to 100.

2. Find two square roots of the number 169

3. Evaluate: 4^2

4. Evaluate: 14^2

5. Find the square root of 400.

(Answers are on page 148.)

ESTIMATING SQUARE ROOTS

For everyday problems, not all measurements come out to be perfect square numbers like 25, 36, and 49. It is important to learn to estimate square roots.

To estimate a square root, you have to think of the perfect squares closest to the number that you have. For example, $\sqrt{90}$ is between what two integers?

Think of the perfect squares closest to 90. $9^2 = 81$ and $10^2 = 100$. Since 90 is between 81 and 100, the square root of 90 must be between 9 and 10.

Let's try another one. Between which two integers does $\sqrt{47}$ lie? We know that $6^2 = 36$, which is too low, and $7^2 = 49$, which is too high. However, 47 is between 36 and 49, so the square root of 47 must be between 6 and 7.

If you want to double-check, get out your calculator!

$\sqrt{36} = 6$ $\quad$ $\sqrt{47} \approx 6.8557$ $\quad$ $\sqrt{49} = 7$

Now let's answer a question about your bedroom. If your floor has an area of 200 square feet, about how much carpet do you need, in feet?

We know that $14^2 = 196$ (too low) and $15^2 = 225$ (too high). Since 200 is between 196 and 225, you need between 14 and 15 feet of carpet.

BRAIN TICKLERS
Set # 40

1. Between which two whole numbers does $\sqrt{80}$ lie?

2. Jackie thinks $\sqrt{29}$ is close to 6. Tim thinks it is close to 7. Who is correct and why?

Each square root is between two numbers. Name the integers.

3. $\sqrt{51}$ 4. $\sqrt{110}$ 5. $\sqrt{140}$

(Answers are on page 148.)

THE REAL NUMBER SYSTEM

In science class, you learn how to classify animals. For example, a gecko is a lizard, which is a reptile, which is an animal. This is an animal system: animal → reptile → lizard → gecko. In math, we classify numbers. The real number system consists of all the points on a number line.

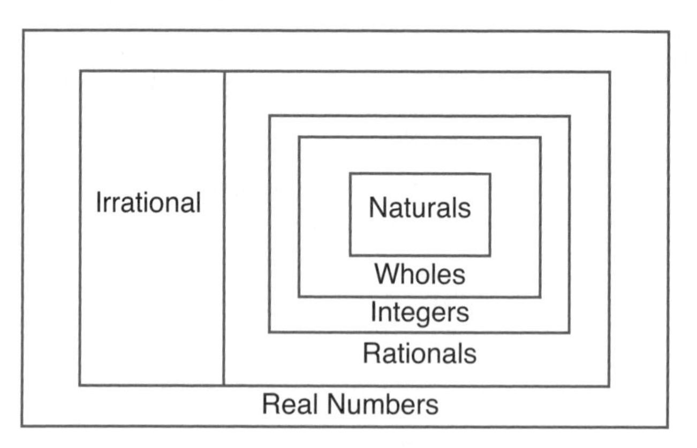

REAL NUMBER SYSTEM

Rational Numbers	Irrational Numbers
A number that can be written as a fraction	A number that cannot be written as a fraction
A decimal that ends/ terminates	A decimal that does not repeat and does not end/ terminate
A decimal that repeats	
Examples	Examples
$\frac{1}{2}$, 0.333 . . . 0.125 $\sqrt{25}$ 1.25	π 0.123456 . . . $\sqrt{46}$

Rational numbers can be broken into more specific families.

NATURAL NUMBERS {1, 2, 3, 4, . . .}

The natural numbers can also be called counting numbers; they are the numbers you count.

WHOLE NUMBERS {0, 1, 2, 3, 4, . . .}
The whole numbers are all of the natural numbers plus zero.

INTEGERS {. . . –3, –2, –1, 0, 1, 2, 3, 4, . . .}
The integers are all of the natural numbers, their opposites, and zero.

To classify or name the sets of numbers that a number belongs to, you name any set that it satisfies.

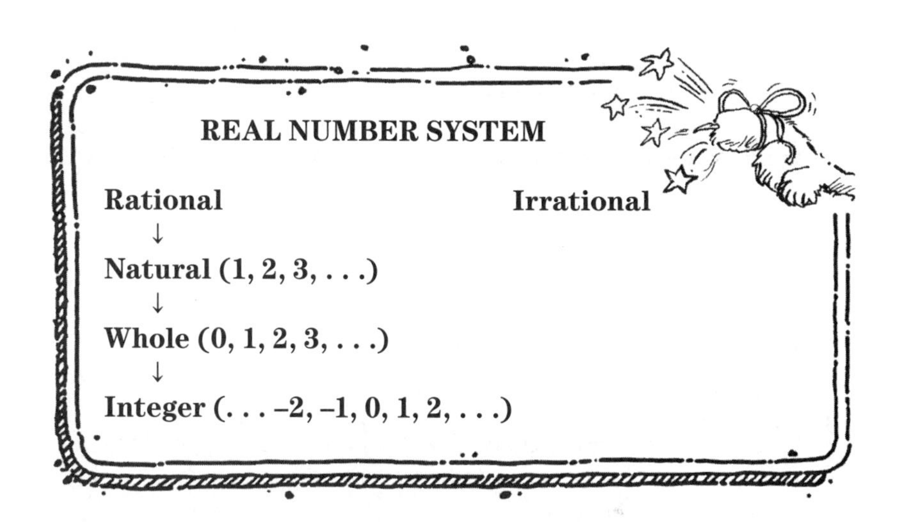

Exciting Examples

Name the sets of numbers that include 5. First you have to decide if 5 is rational or irrational. It is a whole number, so it is rational. 5 is a rational number, a natural number, a whole number, and an integer.

How about –4?
–4 is rational. It is also an integer because it is a negative whole number.

How about $\sqrt{17}$? I cannot figure out the square root of 17 without a calculator. Since it is a nonrepeating decimal that never ends, it is irrational.

How about $\sqrt{100}$? The square root of 100 equals 10. This makes it rational, a natural number, a whole number, and an integer.

Caution—Major Mistake Territory!

A number, like 5, can belong to more than one set of numbers. Think of people. A woman can be a mother, a daughter, a sister, and an aunt. You have to name all of the sets a number belongs to. 5 is a rational number, but it is also a natural number, a whole number, and an integer.

BRAIN TICKLERS
Set # 41

Place a check mark in each appropriate box, to represent the set(s) to which the number belongs.

Example	Rational	Irrational	Integer	Whole	Natural
1. 6					
2. –2					
3. $\sqrt{25}$					
4. 1.25					
5. 0.126498503 . . .					
6. $\frac{3}{4}$					
7. $\sqrt{18}$					
8. 0.55555 . . .					

(Answers are on page 148.)

BRAIN TICKLERS—THE ANSWERS

Set # 35, page 125

1. 8
2. 16
3. 64
4. 1000
5. 1
6. –27
7. 25
8. 1

Set # 36, page 130

1. 5^{10}
2. 7^{15}
3. x^{15}
4. 5^{3}
5. 8^{11}
6. a^{16}
7. 3^{12}
8. 4^{10}
9. x^{15}
10. 10^{10}
11. m^{2}
12. 15^{4}
13. 11
14. 14^{3}
15. b^{7}

Set # 37, page 133

1. $\frac{1}{100,000}$
2. $\frac{1}{81}$
3. $\frac{1}{64}$
4. $\frac{1}{-27}$
5. $\frac{1}{49}$

Set # 38, page 137

1. 24,000,000
2. 36,500,000,000
3. 71,020
4. 0.0052
5. 0.29
6. 0.00000456
7. 2.34×10^{5}
8. 1.123×10^{3}
9. 3.07×10^{8}
10. 5×10^{-4}
11. 3.2×10^{-2}
12. 4.56×10^{-6}

Set # 39, page 141

1. 1, 4, 9, 16, 25, 36, 49, 64, 81, 100
2. 13, –13
3. 16
4. 196
5. 20

Set # 40, page 143

1. 8 and 9
2. Jackie is correct because 29 is closer to 36 than to 49.
3. 7 and 8
4. 10 and 11
5. 11 and 12

Set # 41, page 146

	Example	Rational	Irrational	Integer	Whole	Natural
1.	6	✓		✓	✓	✓
2.	–2	✓		✓		
3.	$\sqrt{25}$	✓		✓	✓	✓
4.	1.25	✓				
5.	0.126498503 . . .		✓			
6.	$\frac{3}{4}$	✓				
7.	$\sqrt{18}$		✓			
8.	0.55555 . . .	✓				

CHAPTER SEVEN

Multistep Equations and Inequalities

TWO-STEP EQUATIONS

Sometimes more than one operation is needed to solve equations. Follow these easy steps to make solving two-step equations painless.

Step 1: Undo addition or subtraction.

Step 2: Undo multiplication or division.

Let's look at a few examples of how to solve two-step equations.

Exciting Examples

Solve: $2x + 3 = 15$

Step 1: Undo addition or subtraction. Since the problem says +3, subtract 3 from both sides of the equation.

$$\begin{array}{rcr} 2x + 3 & = & 15 \\ -3 & & -3 \\ \hline 2x & = & 12 \end{array}$$

Step 2: Undo multiplication or division. Since the problem says "2 times x equals 12," divide both sides by 2.

$$\frac{2x}{2} = \frac{12}{2}$$

$x = 6$

Let's try another one!

Solve: $\frac{x}{3} - 5 = 4$

Step 1: Undo addition or subtraction. Since the problem says –5, add 5 to both sides.

$$\frac{x}{3} - 5 = 4$$
$$\underline{\quad +5 \quad +5}$$
$$\frac{x}{3} \quad = 9$$

Step 2: Undo multiplication or division. The problem reads "x divided by 3 equals 9." To undo division, multiply by 3.

$$3\left(\frac{x}{3}\right) = 9(3)$$
$$x = 27$$

Once you solve an equation, you should check your work to make sure it is correct. To check, just substitute the answer you got for the variable in the original equation. If both sides of the equation are equal, your answer is correct!

Let's check our last problem. We solved $\frac{x}{3} - 5 = 4$ and found $x = 27$.

Step 1: Rewrite the problem.

$$\frac{x}{3} - 5 = 4$$

Step 2: Plug in the answer, 27, for x.

$$\frac{27}{3} - 5 = 4$$

Step 3: Simplify each side using the order of operations

$9 - 5 = 4$
$4 = 4$ Check!

Your equation works, therefore 27 is the correct answer.

BRAIN TICKLERS
Set # 42

Solve these two-step equations.

1. $6x - 3 = 15$
2. $4x + 1 = 9$
3. $-3x + 3 = 15$
4. $\frac{x}{4} + 3 = 10$
5. $\frac{x}{2} - 4 = 8$

(Answers are on page 174.)

COMBINING LIKE TERMS

Before we can solve more equations, we need to take a look at like terms. **Terms** in an expression are separated by plus or minus signs.

$8x + 4y - 3x + 10y$

In the above example there are 4 terms: $8x$, $4y$, $-3x$, and $10y$. When identifying terms, you have to remember that the sign in front of the number goes with the term. Let's look at a few more:

EXPRESSION	# OF TERMS	LIST THE TERMS
$k + 3g - 4k + 10$	4	$k, 3g, -4k, 10$
$7a + 4a - 3b - 4b + 7$	5	$7a, 4a, -3b, -4b, 7$

Like terms such as $8x$ and $5x$ can be grouped together because they have the same variable, raised to the same exponent. Let's look at a few examples.

$3x$ and $4y$	NOT like terms	The variables (letters) are different
$3x$ and 7	NOT like terms	The 7 does not have a matching "x" variable
$4x$ and $3x^2$	NOT like terms	The exponent on the second term, $3x^2$, does not match the exponent on the first term
$3x$ and $4x$	LIKE TERMS!	Both terms have the variable x, raised to the first power
$5x^2$ and $-7x^2$	LIKE TERMS!	Both terms have the variable x raised to the second power

Now that we can identify like terms, we can use this to help make simplifying expressions a lot easier. When trying to combine like terms, a good strategy is to use "Box, Circle, Underline." Box terms that are alike, circle other terms that are alike, and underline other terms that are alike. It's painless, you'll see.

Combine Like Terms

$5x + 3x$

$5x + 3x$ — Circle $5x$ and $3x$ since they are like terms.

$8x$ — Combine the coefficients: $5 + 3 = 8$.

$5a + 7 - a + 12$

$5a + 7 - a + 12$ — Box and circle like terms.

$4a + 19$ — Combine coefficients: $5a - a$, and $7 + 12$.

Caution—Major Mistake Territory!

The coefficient of a variable by itself, such as k, is 1, because $1k = k$.

$7a + 3a + 4b + 5 + 6b$

$7a + 3a + 4b + 5 + 6b$ — Box, circle, and underline like terms.

$10a + 10b + 5$ — Combine coefficients: $7a + 3a$, $4b + 6b$, and 5.

$-9m + 13n - 8m - 6n - 12z$

$-9m + 13n - 8m - 6n - 12z$ — Box, circle, and underline like terms.

$-17m + 7n - 12z$ — Combine coefficients: $-9m - 8m$, $13n - 6n$, $-12z$.

$18x + 3x^2 - 15x - 2x^2 + 2x + 3$

$18x + 3x^2 - 15x - 2x^2 + 2x + 3$ — Box, circle, and underline like terms.

$5x + x^2 + 3$ — Combine coefficients: $18x - 15x + 2x$, $3x^2 - 2x^2$, 3.

BRAIN TICKLERS
Set # 43

Combine like terms.

1. $5x + 12x$
2. $14y - 8 - 10y$
3. $3a + 6 - 2a + 12$
4. $16c + 14d + 10 - 8c - 6d - 2$
5. $7d - d + 5e + 23$

(Answers are on page 174.)

THE DISTRIBUTIVE PROPERTY

Another important step to simplifying expressions and equations is to use the **distributive property**.

What happens when a teacher distributes papers to the class? The teacher will hand out a piece of paper to every student in the class. Well how does this relate to math? Let's look at an example.

$2(6 + 3)$

There are actually two ways to simplify this. The first is to use the order of operations. When using the order of operations, you simplify in the parentheses first. $2(6 + 3) = 2(9) = 18$.

The second way to simplify $2(6 + 3)$ is to use the distributive property. The distributive property allows you to multiply first! The 2 is multiplied (distributed) to each number inside the parentheses, which looks like this:

$2(6 + 3) = 2(6) + 2(3)$
$= 12 + 6$
$= 18$

Remember, the 2 must be distributed (handed out by multiplying) to each term inside the parentheses.

THE DISTRIBUTIVE PROPERTY

To multiply a number by a sum, multiply each number in the sum by the number in front of the parentheses.

$a(b + c) = ab + ac$
$a(b - c) = ab - ac$

Let's try a few! Simplify using the distributive property.

$3(N + 2)$	$4(k - 3)$	$3(3x + 2y)$
$3(N) + 3(2)$	$4(k) - 4(3)$	$3(3x) + 3(2y)$
$3N + 6$	$4k - 12$	$9x + 6y$

Now that you know how to distribute and combine like terms, we can start to do problems like this!

$3(y + 4) + 2y$	
$3(y) + 3(4) + 2y$	Distribute.
$3y + 12 + 2y$	Multiply.
$5y + 12$	Combine like terms.

$5(y - 2) - 10$	
$5(y) - 5(2) - 10$	Distribute.
$5y - 10 - 10$	Multiply.
$5y - 20$	Combine like terms.

MATH TALK!

When simplifying expressions, first you clear the parentheses by using the distributive property, and then you combine like terms.

$$3(x + y) + 12x$$
$$3(x) + 3(y) + 12x$$
$$3x + 3y + 12x$$
$$15x + 12y$$

BRAIN TICKLERS
Set # 44

Simplify.

1. $5(y + 2)$

2. $3(2y + 8)$

3. $4(x - 7)$

4. $3(5x - 1)$

5. $2(3x - 5) + 10$

6. $3(x + 5) + 2x$

(Answers are on page 174.)

MULTISTEP EQUATIONS

You are a pro at solving two-step equations, so multistep equations will be easy! Sometimes, before you can undo addition or subtraction, you might have to simplify the equation. So that will become a new step. These steps will work for solving any equation; just skip step 1 if it is not a multistep equation!

Step 1: Simplify the equation.
→ Do you have to distribute?
→ Do you have to combine like terms?

Step 2: Undo addition or subtraction.

Step 3: Undo multiplication or division.

A Few Examples

Solve: $2x + 5x - 4 = 17$

Step 1: Simplify the equation by combining like terms.
$2x + 5x - 4 = 17$
$7x - 4 = 17$

Step 2: Undo addition or subtraction
$$\begin{array}{rcl} 7x - 4 &=& 17 \\ \underline{\quad +4} && \underline{+4} \\ 7x \quad &=& 21 \end{array}$$

Step 3: Undo multiplication or division
$$\frac{7x}{7} = \frac{21}{7}$$
$x = 3$

Painless! Let's try another!

Solve: $2(x + 4) = 30$

Step 1: Simplify the equation by distributing.
$2(x + 4) = 32$
$2x + 8 = 32$

Step 2: Undo addition or subtraction.

$$\begin{array}{l} 2x + 8 = 32 \\ \quad\ \ -8 \ \ -8 \\ \hline 2x \quad\ \ = 24 \end{array}$$

Step 3: Undo multiplication or division.

$$\frac{2x}{2} = \frac{24}{2}$$

$$x = 12$$

Here is a third example.

Solve: $4(x + 3) + 2x = 48$

Step 1: Simplify by distributing and then combining like terms.

$$4(x + 3) + 2x = 48$$

$$4x + 12 + 2x = 48$$

$$6x + 12 = 48$$

Step 2: Undo addition or subtraction.

$$\begin{array}{l} 6x + 12 = 48 \\ \quad\ \ -12 \ -12 \\ \hline 6x \quad\quad = 36 \end{array}$$

Step 3: Undo multiplication or division.

$$\frac{6x}{6} = \frac{36}{6}$$

$$x = 6$$

Let's check our third example.

Step 1: Rewrite the original problem.

$$4(x + 3) + 2x = 48$$

Step 2: Substitute 6 for x.

$$4(6 + 3) + 2(6) = 48$$

Step 3: Simplify each side using the order of operations.

$$4(9) + 12 = 48$$

$$36 + 12 = 48$$

$48 = 48$ Check!

BRAIN TICKLERS
Set # 45

Solve each of the following equations and check.

1. $3(x - 4) = 24$
2. $4x + 3x + 7x + 5 = 33$
3. $2(x + 1) - 10 = 22$
4. $-3x + 5(x + 1) = 11$

(Answers are on page 174.)

VARIABLES ON BOTH SIDES

Some equations have variables on both sides of the equals sign. Solving an equation with variables on both sides is similar to solving an equation with a variable on only one side. We are used to solving problems like this:

$2x - 5 = 11$

The first step was to undo addition or subtraction. Why? You are combining like terms to get all of the numbers on one side of the equation.

$$\left.\begin{array}{r} 2x - 5 = 11 \\ +5 \quad +5 \end{array}\right\}$$

$$\frac{2x}{2} = \frac{16}{2}$$

$$x = 8$$

For a problem with variables on both sides, you still have to undo addition or subtraction. However, now you also have to move the variables!

Examples

Solve: $2x + 12 = 6x$

In this problem, we have to get all of the x variables on one side of the equation.

Step 1: $2x + 12 = 6x$

Subtract $2x$ from both sides of the equation

$$\begin{array}{r} 2x + 12 = 6x \\ -2x \qquad -2x \\ \hline 12 = 4x \end{array}$$

Step 2: Undo multiplication or division.

$$\frac{12}{4} = \frac{4x}{4}$$

$$3 = x$$

Let's try another one.

Solve: $4y - 8 = 2y + 12$

Step 1: Undo addition or subtraction twice, once for variables and once for numbers.

$$\begin{array}{l} 4y - 8 = 2y + 12 \\ \underline{-2y \quad\;\; -2y} \\ 2y - 8 = 12 \end{array}$$

Helpful Hint!
When variables on are on both sides, move the variables to the left of the equals sign. This will make solving the equation easier.

Undo addition or subtraction again!

$$\begin{array}{l} 2y - 8 = 12 \\ \underline{\quad\;\; +8 \quad +8} \\ 2y \quad\;\; = 20 \end{array}$$

Step 2: Undo multiplication or division.

$$\frac{2y}{2} = \frac{20}{2}$$

$y = 10$

LET'S REVIEW OUR STEPS!

1. **Simplify (distribute and/or combine like terms).**
2. **Undo addition or subtraction (variables and numbers).**
3. **Undo multiplication or division.**

Let's try one more!

Solve: $3(x + 2) + 4 = 2x + 3x + 20$

Step 1: Simplify (distribute, combine like terms).

$3(x + 2) + 4 = 2x + 3x + 20$

$3x + 6 + 4 = 2x + 3x + 20$

Step 2: Undo addition or subtraction.

$3x + 10 = 5x + 20$ → move the variables to the left

$$\begin{array}{l} \underline{-5x \qquad -5x} \\ -2x + 10 = 20 \end{array}$$

$-2x + 10 = 20$ → move the numbers to the right

$$\begin{array}{l} \underline{\qquad -10 \;\; -10} \\ -2x = 10 \end{array}$$

Step 3: Undo multiplication or division.

$$\frac{-2x}{-2} = \frac{10}{-2}$$

$x = -5$

BRAIN TICKLERS
Set # 46

Solve.

1. $8x - 2 = 6x + 10$
2. $2x + 6x - 4 = 4x + 24$
3. $-2(x + 1) = 4x - 8$
4. $4(x - 3) = 5(x + 3)$
5. $5x + 3 = 14 - 6x$

(Answers are on page 174.)

INEQUALITIES

An **inequality** is a mathematical sentence that compares two quantities using $>$, $<$, $\geq$, or $\leq$.

Let's review the symbols used for writing inequalities.

SYMBOL $<$	SYMBOL $>$
Is less than Is fewer than	Is greater than Is more than Exceeds

SYMBOL $\leq$	SYMBOL $\geq$
Is less than or equal to Is no more than Is at most	Is greater than or equal to Is at least Is no less than

Here are some common phrases you might find in everyday life that involve inequalities.

Phrase	**Inequality**
You must be **at least 16 years or older** to drive	$a \geq 16$
The **maximum capacity** in the auditorium **is 700** people	$c \leq 700$
Kelly owns **more than 200** DVDs	$k > 200$
Tracy has **fewer than 20 pairs of jeans**.	$t < 20$

An inequality that contains one or more variable is called an **algebraic inequality**. A number that makes the inequality true is called the **solution.** The solution set of an inequality can be shown by graphing it on a number line.

To graph an inequality, use these painless steps.

Step 1: Locate the number in the inequality on the number line.

Step 2: If it is $>$ or $<$, place an open circle over the number. ○
If it is $\geq$ or $\leq$, place a closed circle over the number. ●

Step 3: For greater than or greater than or equal to, shade to the right.
For less than or less than or equal to, shade to the left.

Word Phrase	**Inequality**
x is less than 5	$x < 5$ 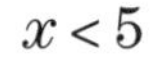

Solution Set

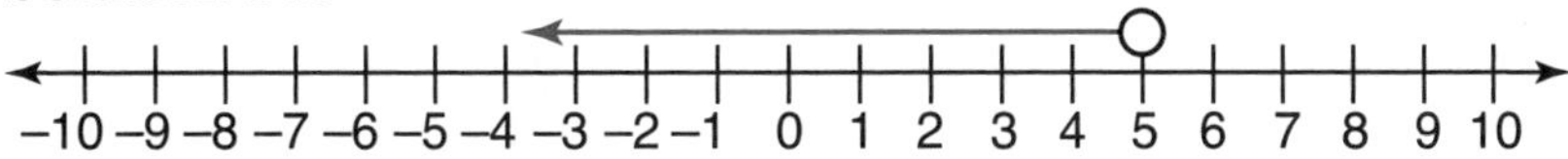

a is greater than –2 $a > -2$

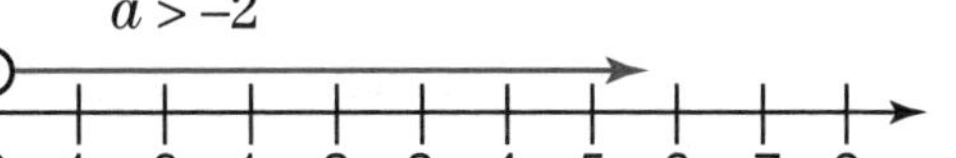

c is less than or equal to 4
c is at most 4 $c \leq 4$ 

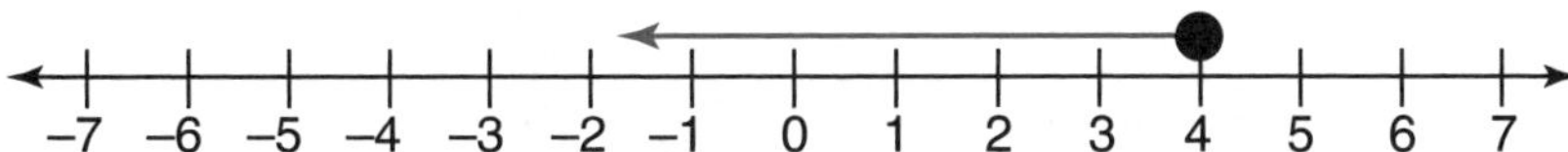

w is greater than or equal to 3
w is at least 3 $w \geq 3$

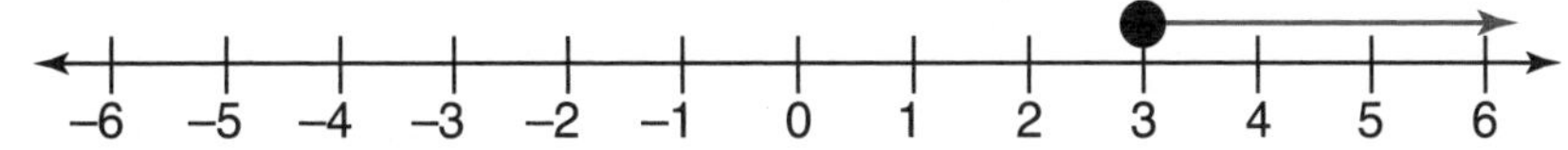

Caution—Major Mistake Territory!

When reading or graphing inequalities, always make sure the variable comes first. For example:

$3 < x$ is the same as $x > 3$.

The inequality is easier to read and understand if the variable comes first!

BRAIN TICKLERS
Set # 47

Graph the following inequalities.

1. $x > 4$

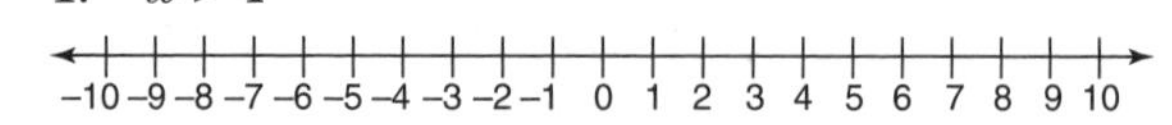

2. $x < 2$

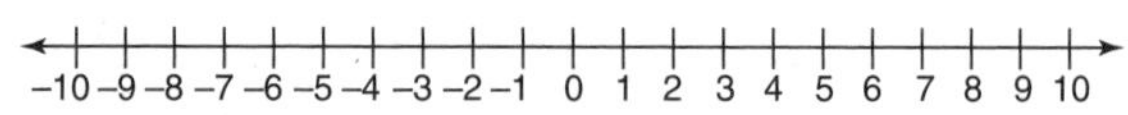

3. $x \geq -4$

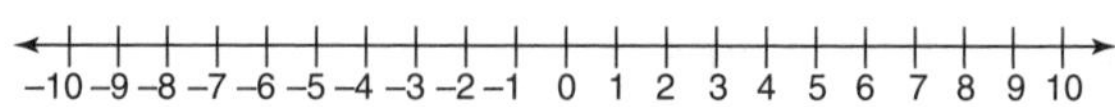

4. $x \leq 0$

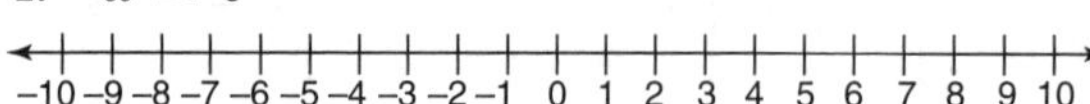

5. $x < 1$

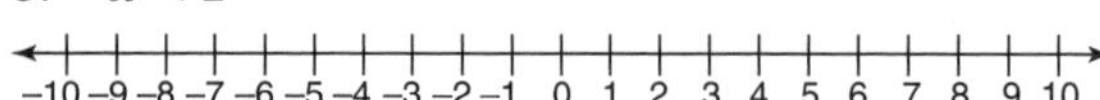

(Answers are on page 174.)

ONE-STEP INEQUALITIES BY ADDING OR SUBTRACTING

Solving inequalities is almost exactly the same as solving an equation. When you solve an equation, you get an answer like $x = 4$. When you solve an inequality, you get an answer like $x > 4$. If $x > 4$, many answers make that statement true. Therefore, you can represent the solutions on a number line.

Let's review our steps for solving an equation.

Step 1: Simplify (distribute, combine like terms).

Step 2: Undo addition or subtraction.

Step 3: Undo multiplication or division.

When solving an inequality, we have to add one more step.

Step 4: Graph the solution on a number line.

Examples

Solve: $x + 3 > 10$

This is a one-step inequality. We only have to undo addition or subtraction.

$$\begin{array}{l} x + 3 > 10 \\ \quad -3 \quad -3 \\ \hline x \qquad > 7 \end{array}$$

Now we can graph "x is greater than 7" on a number line.

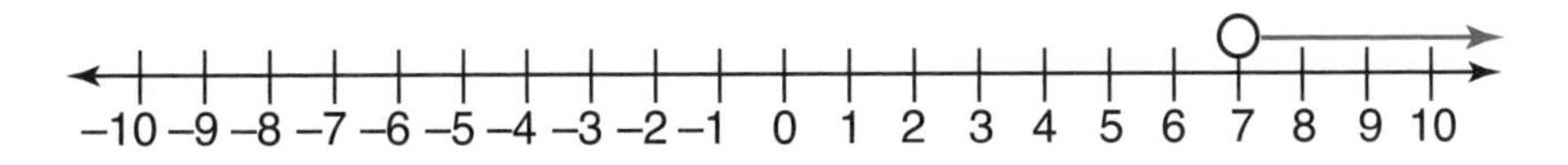

Put an open circle over the number 7. The circle is open to indicate 7 is not included in the answer. Since x is greater than 7, shade to the right of 7.

Solve: $x - 4 \leq -3$

Undo addition or subtraction.

$$\begin{array}{rcr} x - 4 & \leq & -3 \\ +4 & & +4 \\ \hline x & \leq & 1 \end{array}$$

Now graph $x \leq 1$ on a number line. Since x can be equal to 1, put a closed circle over 1 to show that it is included in the answer. Since x is less than or equal to, shade to the left of 1.

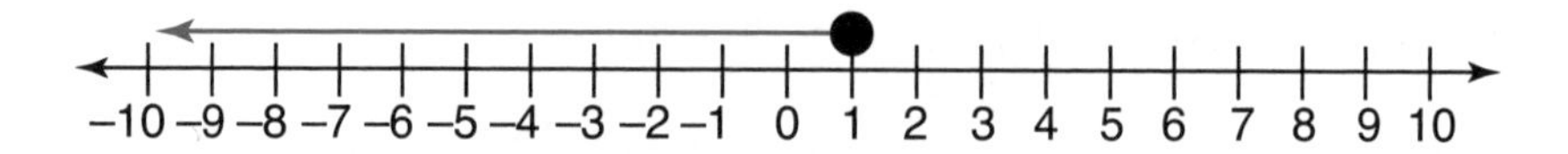

BRAIN TICKLERS
Set # 48

Solve and graph each inequality.

1. $x - 5 > 3$

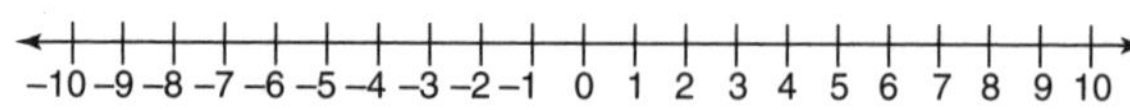

2. $x + 3 < -4$

–10 –9 –8 –7 –6 –5 –4 –3 –2 –1 0 1 2 3 4 5 6 7 8 9 10

3. $x + 8 \geq 10$

–10 –9 –8 –7 –6 –5 –4 –3 –2 –1 0 1 2 3 4 5 6 7 8 9 10

4. $x - 6 \leq -2$

–10 –9 –8 –7 –6 –5 –4 –3 –2 –1 0 1 2 3 4 5 6 7 8 9 10

(Answers are on page 175.)

ONE-STEP INEQUALITIES BY MULTIPLYING OR DIVIDING

The steps for solving inequalities by multiplying or dividing are the same as those for solving equations. There is one extra rule to remember. If you multiply or divide the variable by a negative number, the inequality sign must be reversed. Let's try a few.

Examples

Solve and graph $2x > 4$.

Since this is just a one-step inequality, undo the multiplication by dividing each side by 2.

$\frac{2x}{2} > \frac{4}{2}$

$x > 2$

When graphing, use an open circle since 2 is not included in the answer. Shade to the right since x is greater than 2.

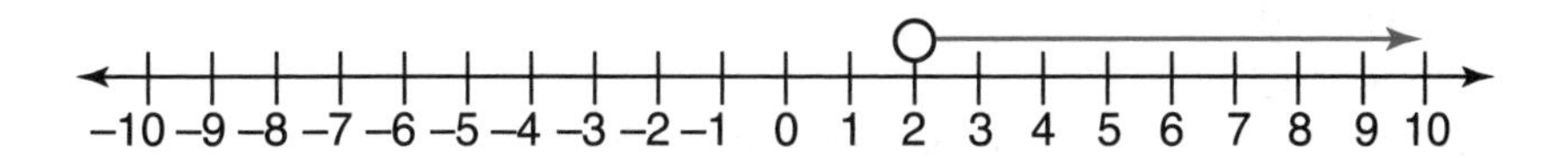

Solve and graph $-3x \leq 9$.

Undo the multiplication by dividing each side by –3. Since you are dividing by a negative number, you must flip the inequality sign!

$-3x \leq 9$

$\frac{-3x}{-3} \geq \frac{9}{-3}$ → Dividing by –3 changes $\leq$ to $\geq$.

$x \geq -3$

Put a closed circle on –3 since it is included in the answer. Shade to the right.

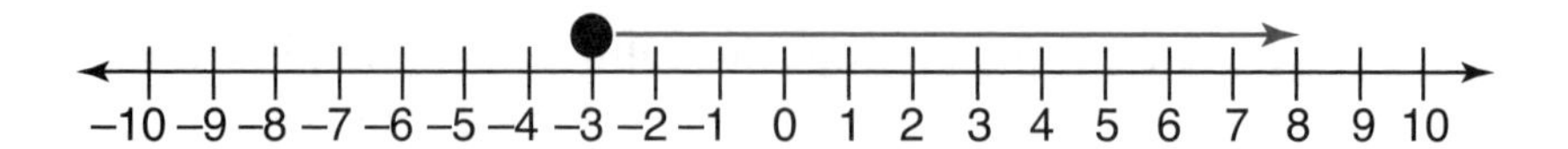

Solve and graph $\frac{x}{3} < 2$.

To undo division, multiply each side by 3.

$3\left(\frac{x}{3}\right) < 2(3)$

$x < 6$

Put an open circle on 6 and shade to the left.

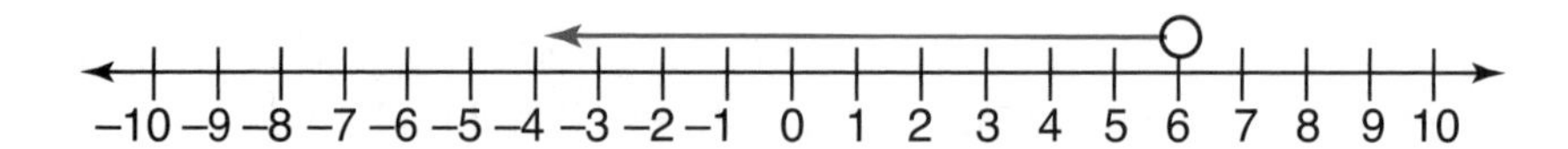

Solve and graph $\frac{x}{-2} \geq 2$.

Undo the division by multiplying both sides by –2. Since you are multiplying the variable by a negative, you must flip the inequality sign!

$\frac{x}{-2} \geq 2$

$-2\left(\frac{x}{-2}\right) \leq 2(-2)$ → Flip the inequality sign here!

$x \leq -4$

Put a closed circle on –4 and shade to the left.

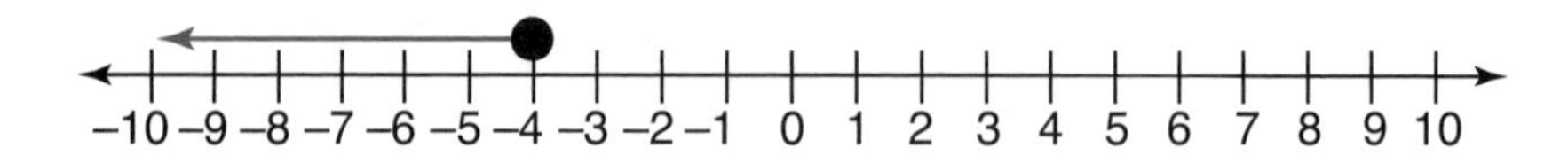

Caution—Major Mistake Territory!

When you multiply or divide the variable by a negative number, you must flip the inequality sign!

$-2x > -4$

Divide both sides by –2, so flip the sign!

$$\frac{-2x}{2} < \frac{-4}{-2}$$

$x < 2$

BRAIN TICKLERS
Set # 49

Solve and graph the inequalities.

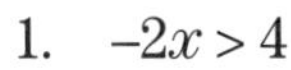

1. $-2x > 4$

–10 –9 –8 –7 –6 –5 –4 –3 –2 –1 0 1 2 3 4 5 6 7 8 9 10

2. $4x \leq 12$

–10 –9 –8 –7 –6 –5 –4 –3 –2 –1 0 1 2 3 4 5 6 7 8 9 10

3. $\dfrac{x}{4} \geq 1$

–10 –9 –8 –7 –6 –5 –4 –3 –2 –1 0 1 2 3 4 5 6 7 8 9 10

4. $\dfrac{x}{-3} < 2$

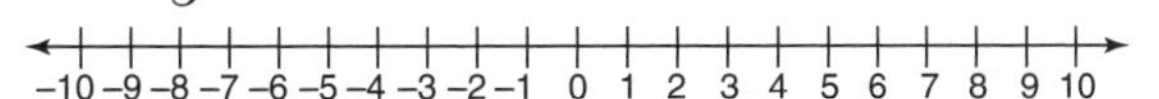

(Answers are on page 175.)

BRAIN TICKLERS—THE ANSWERS

Set # 42, page 153

1. $x = 3$
2. $x = 2$
3. $x = -4$
4. $x = 28$
5. $x = 24$

Set # 43, page 156

1. $17x$
2. $4y - 8$
3. $a + 18$
4. $8c + 8d + 8$
5. $6d + 5e + 23$

Set # 44, page 159

1. $5y + 10$
2. $6y + 24$
3. $4x - 28$
4. $15x - 3$
5. $6x$
6. $5x + 5$

Set # 45, page 162

1. $x = 12$
2. $x = 2$
3. $x = 15$
4. $x = 3$

Set # 46, page 165

1. $x = 6$
2. $x = 7$
3. $x = 1$
4. $x = -27$
5. $x = 1$

Set # 47, page 168

1. $x > 4$

2. $x < 2$

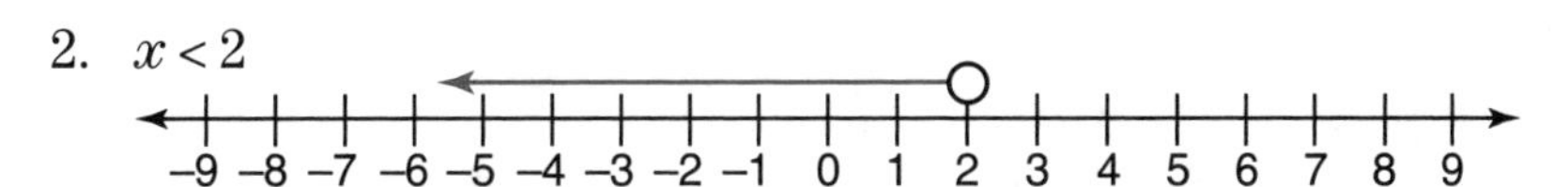

3. $x \geq -4$

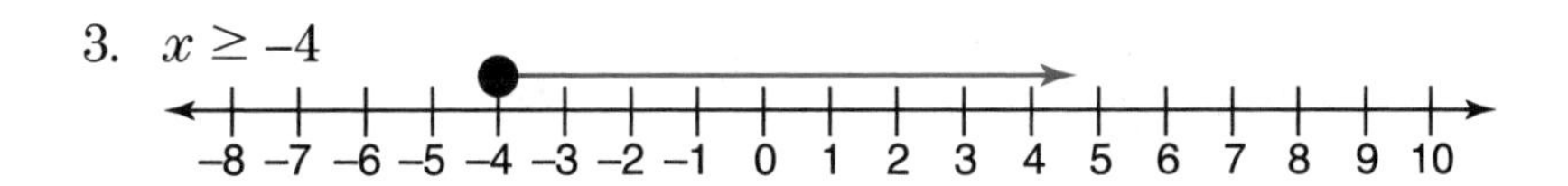

4. $x \leq 0$

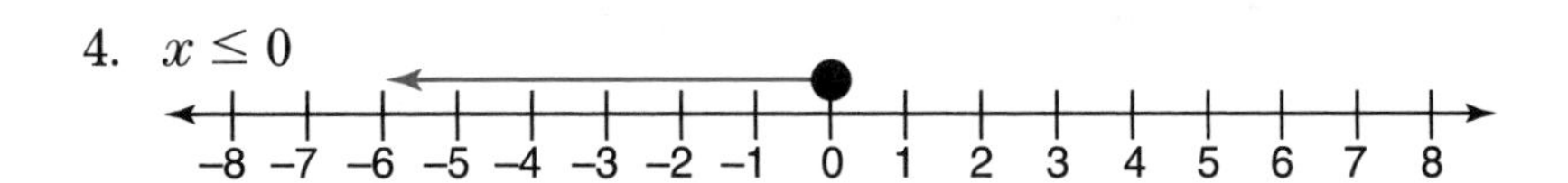

5. $x < 1$

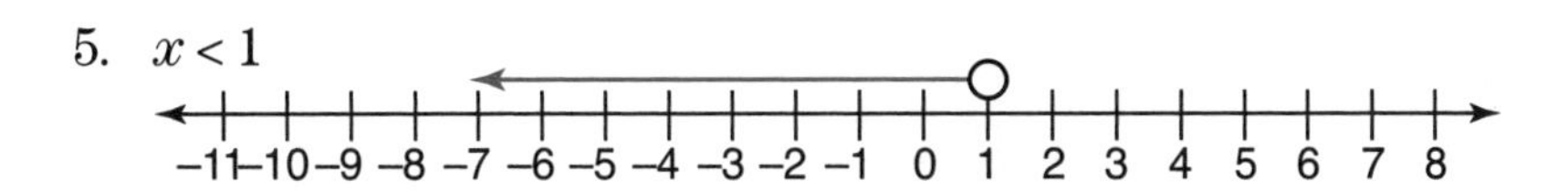

Set # 48, page 170

1. $x > 8$

2. $x < -7$

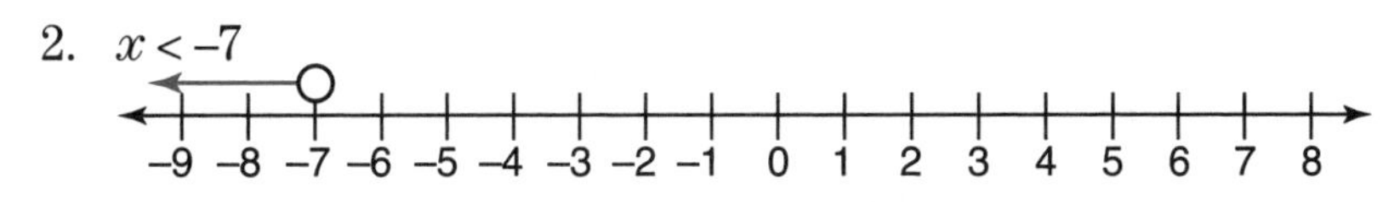

3. $x \geq 2$

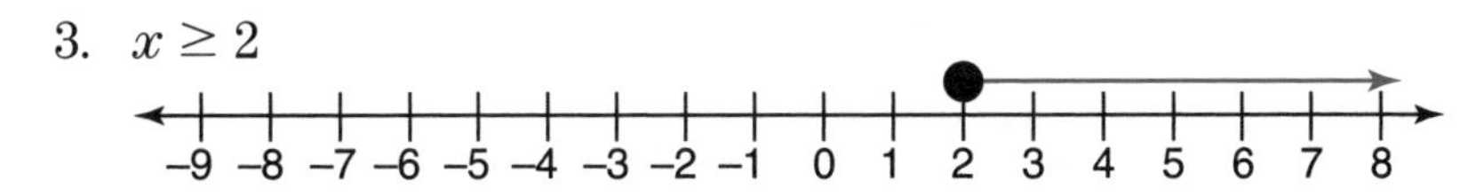

4. $x \leq 4$

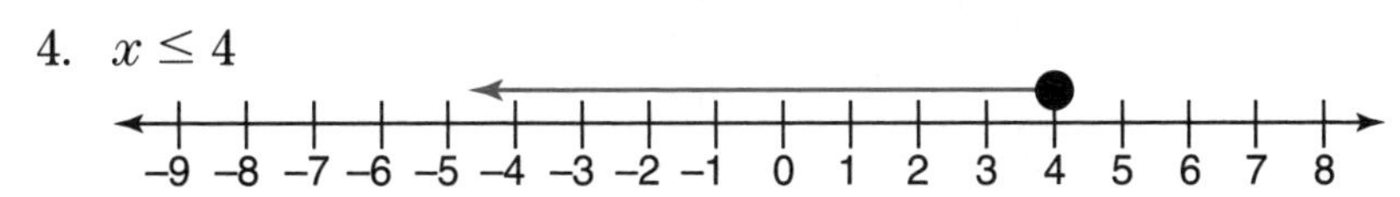

Set #49, page 173

1. $x < -2$

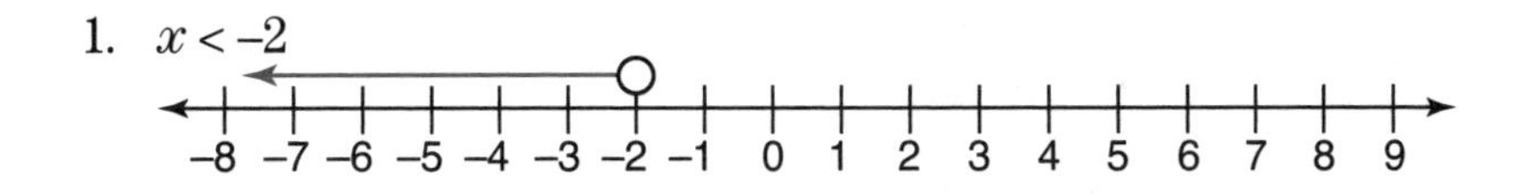

2. $x \leq 3$

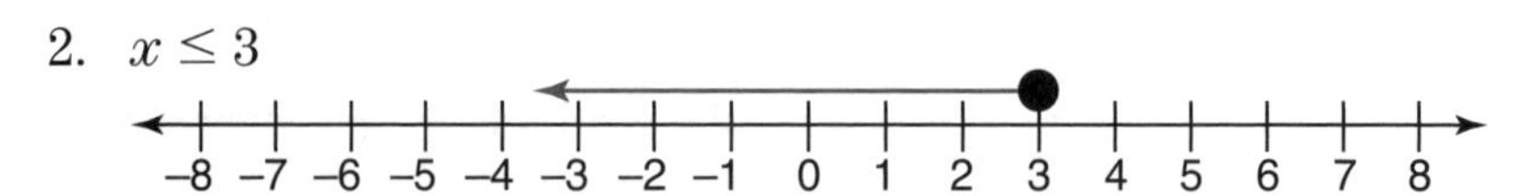

3. $x \geq 4$

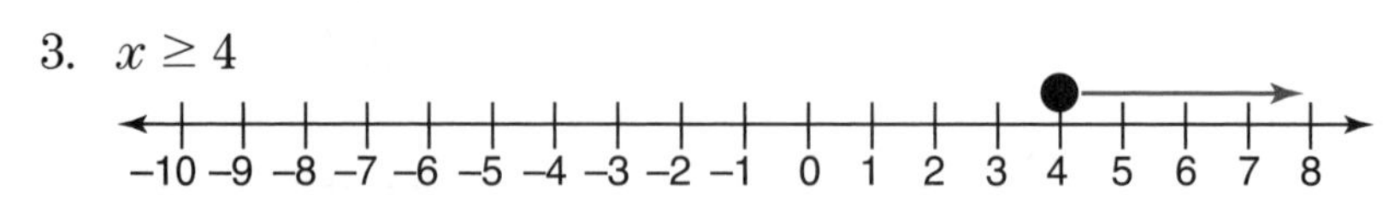

4. $x > -6$

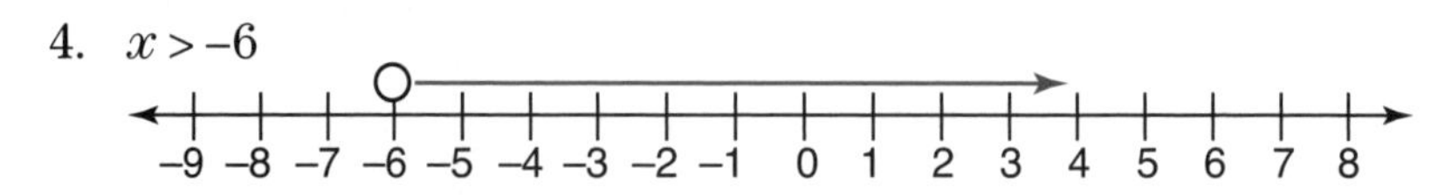

CHAPTER EIGHT
Linear Equations
WHICH WAY TO THE BUNNY SLOPE?
3/2
0
1/2

GRAPHING A LINE USING A TABLE OF VALUES

If you can graph points, you can graph a straight line. Setting up an x-y chart will make graphing painless. Follow these easy steps.

Step 1: Solve the equation for y.

Step 2: Choose three values for x to help you make your chart.

Step 3: Substitute each value for x, to find the value of y.

Step 4: Graph the three points (x, y)

Step 5: Connect these three points with a straight line, and put arrows on the line.

MATH TALK!

When choosing x-values, choose a negative, zero, and a positive. This will cover every section of the graph paper.

Make sure you extend the line past the three points, and put arrows on the ends. This indicates the line goes on forever.

Example 1

Graph the line $y = x + 3$.

Step 1: The line is already in the form $y =$.

Step 2: Choose three values for x.

Step 3: Substitute each value for x and solve for y.

x	$x + 3$	y
–2	–2 + 3	1
0	0 + 3	3
2	2 + 3	5

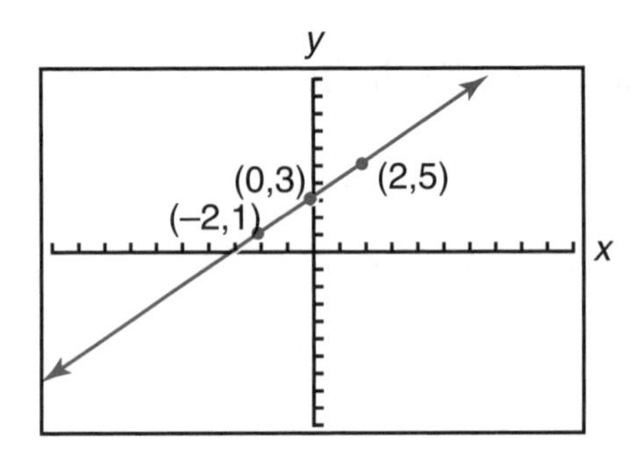

Step 4: Plot the points (–2,1), (0,3) and (2,5).

Step 5: Connect the three points with a straight line, and put arrows on the line.

MATH TALK!

When using a table of values to graph lines, it is not necessary to write the equation in $y =$ form (function form). However, this form does make graphing the line easier.

Example 2

Graph $y + x = 3$.

Step 1: Solve the equation for $y =$. Add x to both sides

$$\begin{array}{l} y - x = 5 \\ \underline{+x \quad +x} \\ y = 5 + x \text{ (or } y = x + 5) \end{array}$$

Step 2: Choose three values for x.

Step 3: Substitute each value for x and solve for y.

x	$5 + x$	y
–3	5 + –3	2
0	5 + 0	5
3	5 + 3	8

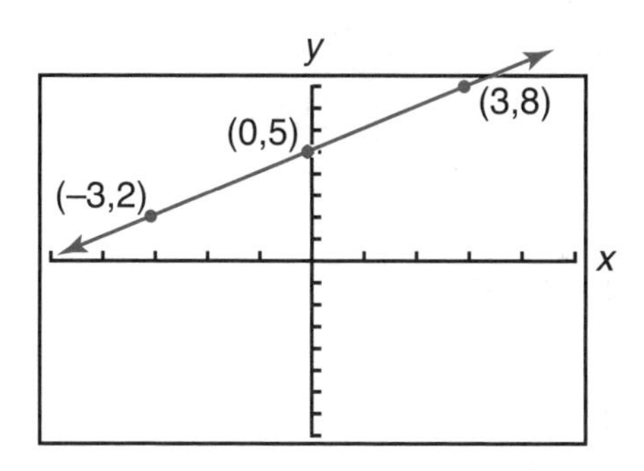

Step 4: Plot the points (–3,2), (0,5), and (3,8).

Step 5: Connect the three points with a straight line, and put arrows on the line.

Example 3

Graph $6x + 2y = 12$.

Step 1: Solve the equation for $y =$
→ Subtract $6x$ from both sides.

$$\begin{array}{l} 6x + 2y = 12 \\ \underline{-6x \qquad -6x} \\ \dfrac{2y}{2} = \dfrac{-6x}{2} + \dfrac{12}{2} \end{array}$$

→ Since y has a coefficient of 2, divide by 2 to get y alone. Both −6 and 12 must also be divided by 2. $y = -3x + 6$

Step 2: Choose three values for x.

Step 3: Substitute each value for x and solve for y.

x	$-3x + 6$	y
−2	$-3(-2) + 6$	12
0	$-3(0) + 6$	6
2	$-3(2) + 6$	0

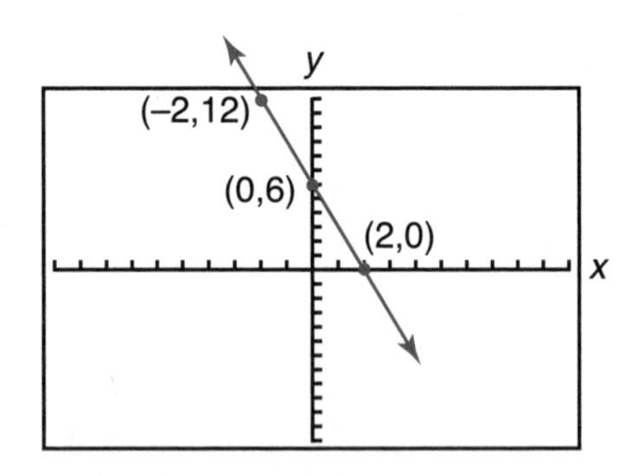

Step 4: Plot the points (−2,12), (0,6), and (2,0).

Step 5: Connect the three points with a straight line, and put arrows on the line.

Caution—Major Mistake Territory!

When solving for y, if y has a number in front of it (coefficient), when you divide y by the number to get y alone, you must also divide EACH term on the other side of the equation by that number!

Example:
Given: $3y = 12x + 15$
To solve for y, you must divide by 3; therefore divide EACH term by 3.

$$\frac{3y}{3} = \frac{12x}{3} + \frac{15}{3}$$

$y = 4x + 5$

BRAIN TICKLERS
Set # 50

If needed, solve each equation for *y*. Make a table of values, and graph the line.

1. $y = 2x + 1$
2. $x + y = -2$
3. $y - 5 = x$
4. $2x - y = 3$

(Answers are on page 217.)

GRAPHING HORIZONTAL AND VERTICAL LINES

Picture the horizon or the floor that you walk on. These are examples of horizontal lines. Horizontal lines go from left to right and are **parallel to the x-axis**. Horizontal lines are written in the form $y = b$, where b is where the line crosses the y-axis. Examples of equations of horizontal lines are $y = 2$, $y = -3$, and $y = 0$. All points on the horizontal line have the same y-coordinate.

The walls in your house or a telephone pole represent vertical lines. Vertical lines go straight up and down and are parallel to the y-axis. A vertical line is written in the form $x = a$, where a is where the line crosses the x-axis. Examples of equations of vertical lines are $x = 2$, $x = -4$, and $x = 0$.

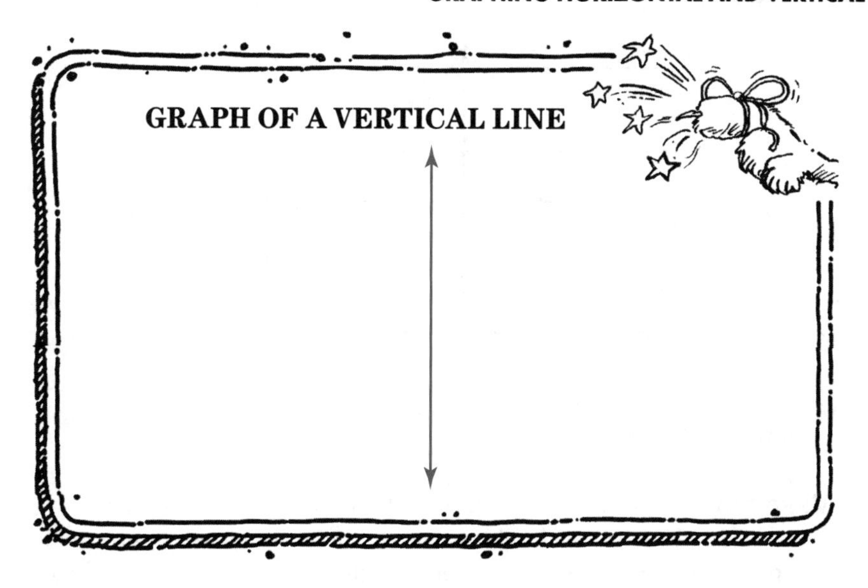

To graph horizontal and vertical lines, make a table of values.

Examples

Graph $y = 2$. For a table, this means the y-value always has to be 2. Choose three different x-values to make your table. No matter what x values you choose, the y-value will always be 2. Then plot the points, and draw the line through the points. This is a horizontal line.

x	y
–3	2
0	2
3	2

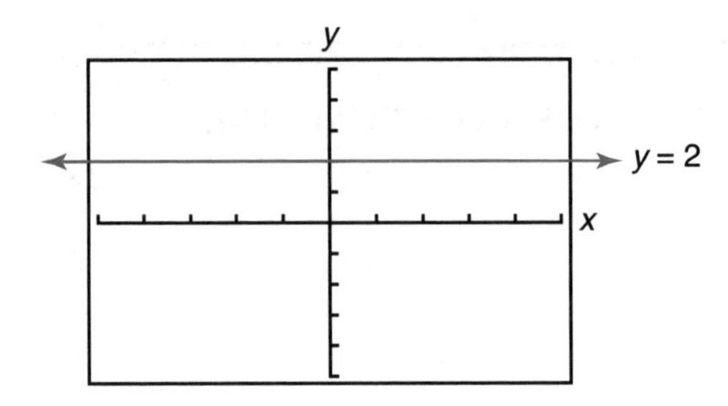

Graph $x = -1$.

Make a table, but this time the x-values always stay the same. The x-value has to be –1. Choose three different y-values to help you make the graph. Plot the points, and draw the line. This is a vertical line.

x	y
–1	–2
–1	0
–1	2

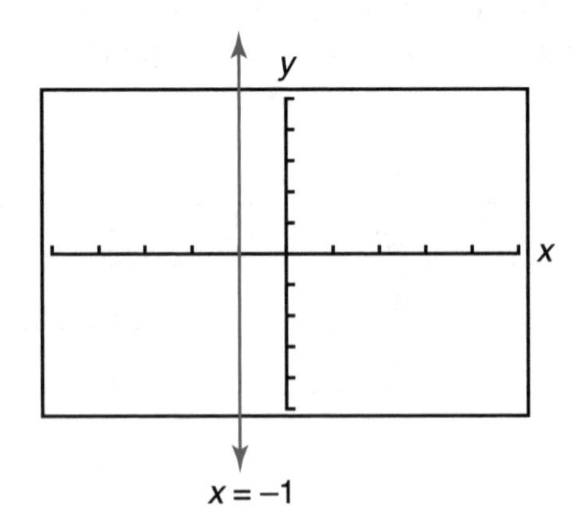

Caution—Major Mistake Territory!

A horizontal line is always written in the form $y = b$. A vertical line is always written in the form $x = a$.

Examples:
$y = 3$ → Horizontal line

$x = -2$ → Vertical line

BRAIN TICKLERS
Set # 51

Graph each line.

1. $x = 4$ 2. $x = -2$ 3. $y = 5$
4. $y = -1$

(Answers are on page 219.)

X- AND Y-INTERCEPTS

The ***x*-intercept** of a graph is where the graph crosses the x-axis. This graph crosses the x-axis at (6,0), so the x-intercept is 6.

The ***y*-intercept** of a graph is where the graph crosses the y-axis. This graph crosses the y-axis at (0,3), so the y-intercept is 3.

You can use the x- and y-intercepts to help you graph a line.

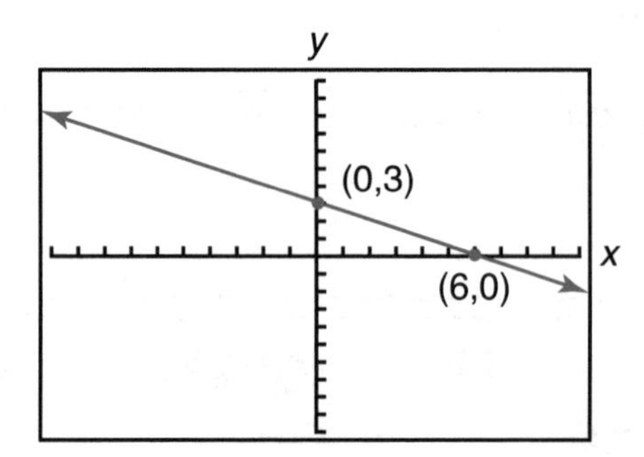

What if the problem says, "Graph the line with an x-intercept of –4 and a y-intercept of 2"? No problem!

The x-intercept is –4, so plot the coordinate (–4,0) since it is on the x-axis. The y-intercept is 2, so plot the coordinate (0,2) since it is on the y-axis. Then draw a line through the two points!

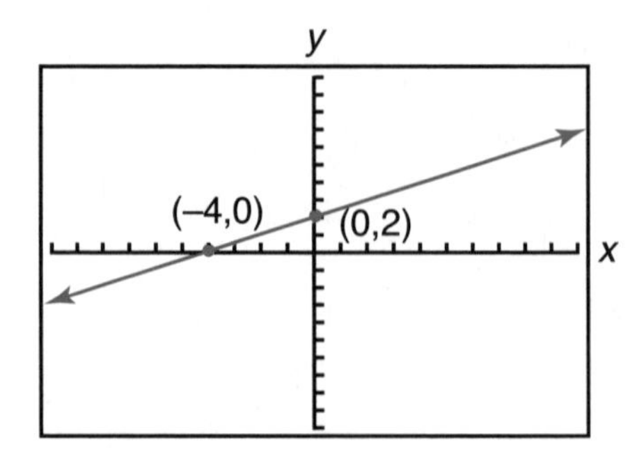

If you already have the equation of a line and want to find the x- and y-intercepts, you need to know two big things!

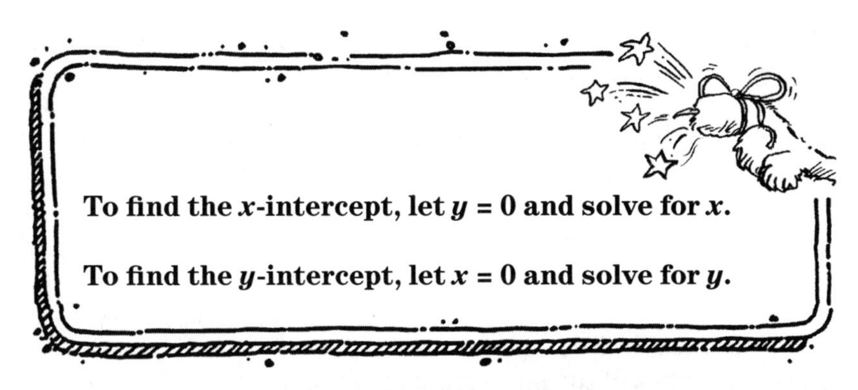

Example 1

Find the x- and y-intercepts of $2x + y = 10$.

x-intercept: Let $y = 0$, and solve for x.

$$2x + 0 = 10$$
$$\frac{2x}{2} = \frac{10}{2}$$
$$x = 5$$

The x-intercept is 5.
The coordinates of the x-intercept are (5,0).

y-intercept: Let $x = 0$, and solve for y.

$$2(0) + y = 10$$
$$0 + y = 10$$
$$y = 10$$

The y-intercept is 10.
The coordinates of the y-intercept are (0,10).

Example 2

Find the x- and y-intercepts of $-3x + 6y = 18$.

x-intercept: Let $y = 0$, and solve for x.

$$-3x + 6(0) = 18$$
$$\frac{-3x}{-3} = \frac{18}{-3}$$
$$x = -6$$

The x-intercept is –6.
The coordinates of the x-intercept are (–6,0).

y-intercept: Let $x = 0$, and solve for y.

$$-3(0) + 6y = 18$$
$$\frac{6y}{6} = \frac{18}{6}$$
$$y=3$$

The y-intercept is 3.
The coordinates of the y-intercept are (0,3).

You can use the x- and y-intercepts to graph the lines. Plot the intercepts (–6,0) and (0,3). Then draw a line connecting the 2 points. Label the line with the equation.

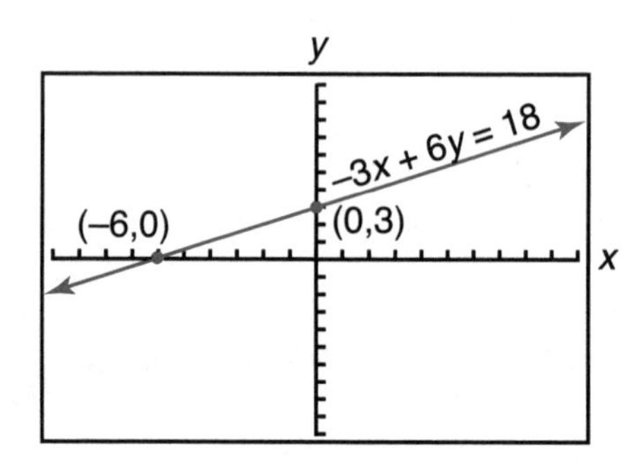

BRAIN TICKLERS
Set # 52

Find the x- and y-intercepts.

1. $2x + 4y = 20$
2. $2x - 10 = 2y$
3. $y = 2x + 8$

Find the x- and y-intercepts. Graph each line using the intercepts.

4. $x + 2y = 6$
5. $3x = 3y - 9$

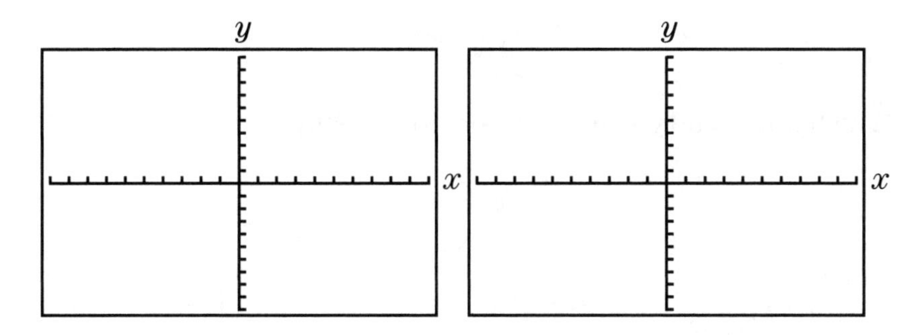

(Answers are on page 220.)

SLOPE

Which hill would you prefer to ski down? Why?

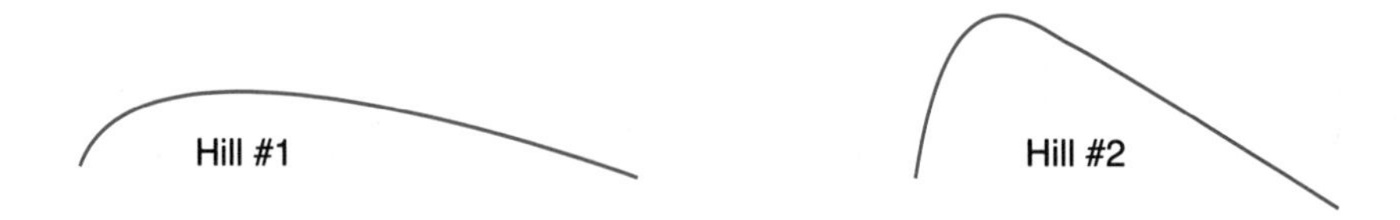

In math, the measure of the steepness of the hill is called **slope.** Notice that hill #1 is less steep than hill #2. What does slope mean in math?

DEFINITION

Slope is a measure of the slant of a line.

TRICK TO REMEMBER

$$\text{Slope} = \frac{\text{Rise (change in } y)}{\text{Run (change in } x)}$$

EXAMPLES

positive slope slants upward, from left to right.

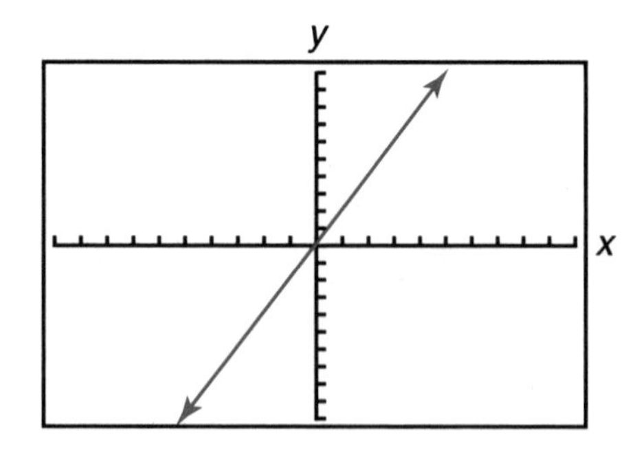

negative slope slants downward, from left to right.

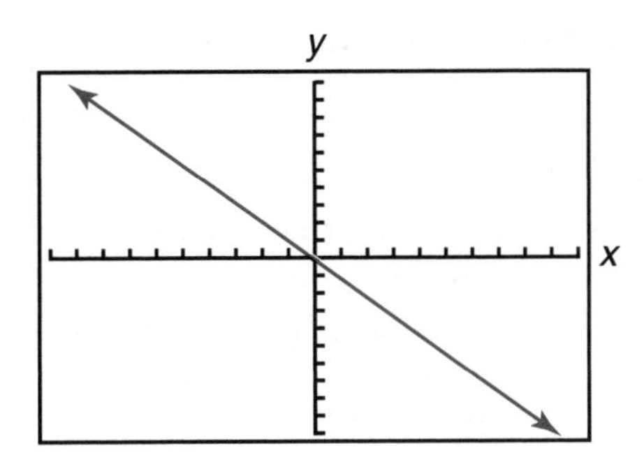

SPECIAL CASES

Horizontal lines have a **zero slope**. The line is horizontal like a floor.

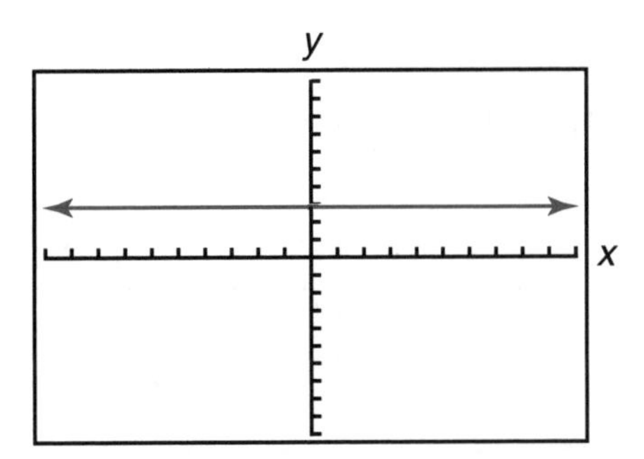

Vertical lines have an **undefined slope**. The line is up and down.

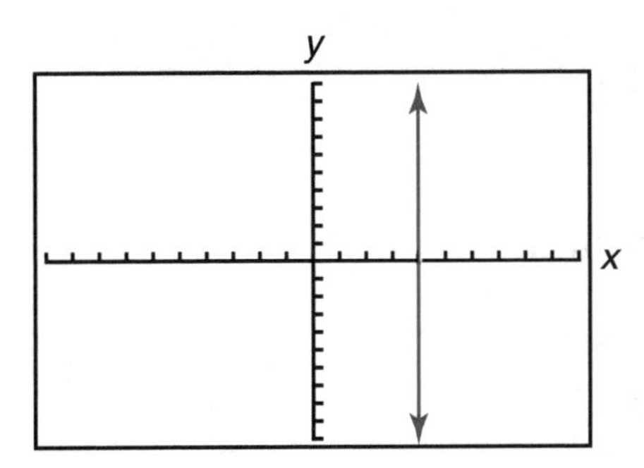

To find the slope of a line, use the phrase "rise over run."

$$\frac{\text{Rise} \updownarrow}{\text{Run} \leftrightarrow}$$

- **Rise** means the vertical distance between points. Rise involves the y-axis. A positive rise means move up. A negative rise means move down.
- **Run** means the horizontal distance between points. Run involves the x-axis. A positive run means move right. A negative run means move left.

You can find the slope of a line using two different methods. One method is to graph the two points and find the slope using rise over run. The second method is to find the slope using the formula

$$m = \frac{y_2 - y_1}{x_2 - x_1}.$$

METHOD 1

Find the slope of the following lines.

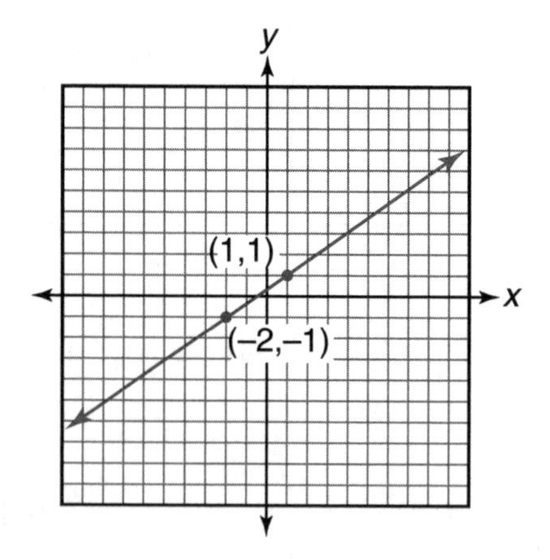

What does slope mean? $\frac{\text{Rise}}{\text{Run}}$

Start at the lowest point, (–2,–1).
Rise up 2 (positive).
Run right 3 (positive).

Slope = $\frac{2}{3}$.

MATH TALK!

If you always start with the lowest point, you will rise in a positive direction every time.

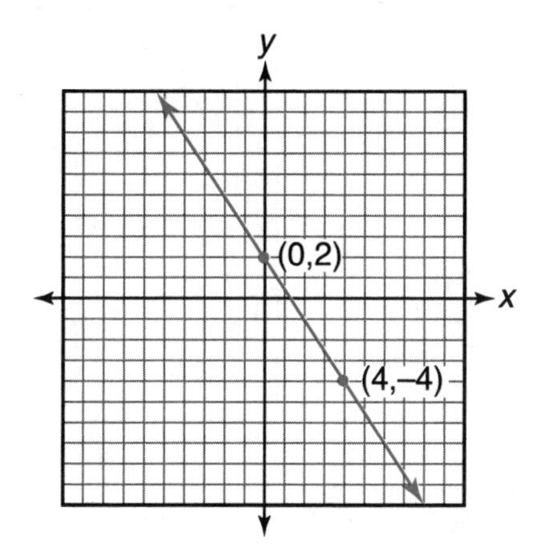

Slope = $\frac{\text{Rise}}{\text{Run}}$

Start at the lowest point (4,–4).
Rise up 6 (positive).
Run left 4 (negative).

Slope = $\frac{6}{-4}$

The slope can reduce to $\frac{3}{-2}$, or $-\frac{3}{2}$.

Caution—Major Mistake Territory!

Reducing the slope does not change your answer. You will still plot a point on the same line!

Slope = $\frac{4}{6}$. Slope = $\frac{2}{3}$.

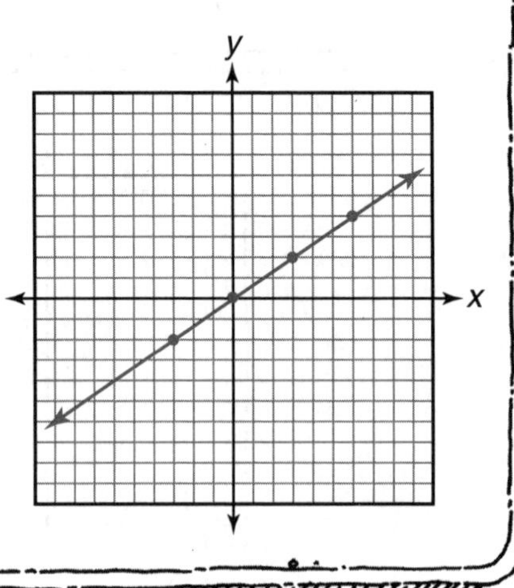

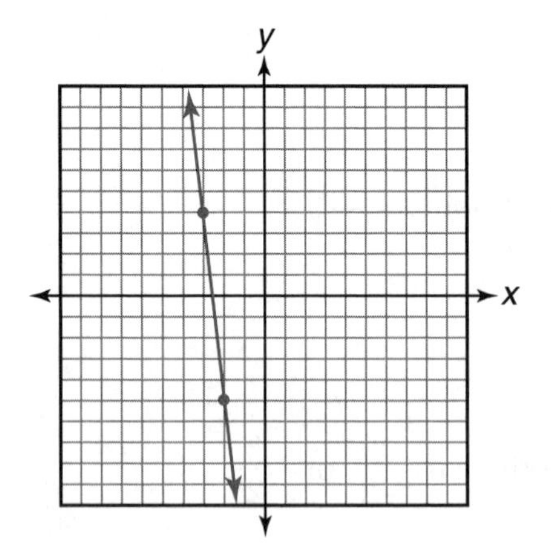

Start at the lowest point (–2,–5).
Rise up 9 (positive) to get to the other point (–2,–5).
Run left 1 (negative) to get to the other point (–2,–5).

Slope = $\frac{9}{-1}$

Slope = -9

MATH TALK!

$\frac{9}{-1}$ is the same as $\frac{-9}{1}$

Moving up 9 and left 1 from the lowest point will get you to the same place as moving down 9 and right 1 when starting from the highest point.

METHOD 2

To find the slope of a line when given two points, use the formula $m = \frac{\text{rise}}{\text{run}} = \frac{y_2 - y_1}{x_2 - x_1}$.

What this formula means is that you find the difference in the y-coordinates (subtract the y-values) and divide by the difference of the x-coordinates (subtract the x-values).

It is important to remember that for the coordinate (5,6), 5 is the x-coordinate and 6 is the y-coordinate.

Find the slope of a line passing through the points (4,3) and (5,6).

$$m = \frac{y_2 - y_1}{x_2 - x_1}$$

$$m = \frac{6 - 3}{5 - 4}$$

$$m = \frac{3}{1}$$

The slope equals 3.

Always have the y-values in the numerator. Also, you have to subtract the x-values in the same order that you subtracted the y-values. What does this mean? Whichever y-value comes first on top, the x value from the SAME COORDINATE PAIR has to come first as well, for the bottom of the fraction.

Example 1

Find the slope of a line passing through the points (–2,4) and (5,6).

$$m = \frac{y_2 - y_1}{x_2 - x_1}$$

$$m = \frac{4 - 6}{-2 - 5}$$

$$m = \frac{-2}{-7}$$

The slope equals $\frac{2}{7}$.

Let's try one more to make sure we have it!

Example 2

Find the slope of a line passing through the points (4,–3) and (–5,2).

$$m = \frac{y_2 - y_1}{x_2 - x_1}$$

$$m = \frac{-3 - 2}{4 - (-5)}$$

$$m = \frac{-5}{9}$$

$$m = -\frac{5}{9}$$

Caution—Major Mistake Territory!

When using the slope formula, make sure you always have the y-values on top!

(5,7) and (4,3)

$$m = \frac{y_2 - y_1}{x_2 - x_1} = \frac{7 - 3}{5 - 4} = \frac{4}{1} = 4$$

Slope—Two special cases

When graphing lines, some lines slant up and have a positive slope ↗ and some lines slant down and have a negative slope ↘.

But what happens if the lines are either horizontal or vertical?

Horizontal Lines

A horizontal line does not slant up or down. Therefore it does not have a positive or a negative slope. **A horizontal line has a slope of zero**.

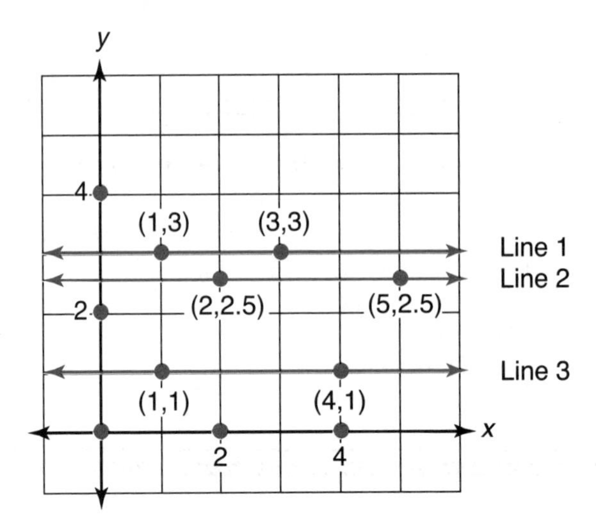

Let's prove that a horizontal line has a slope of zero by using our slope formula.

Let's use the coordinates of Line 1: (1,3) and (3,3).

$$m = \frac{y_2 - y_1}{x_2 - x_1}$$

$$m = \frac{3 - 3}{1 - 3}$$

$m = \frac{0}{-2}$, which equals 0!

Now use the coordinates of Line 2: (2,2.5) and (5,2.5).

$$m = \frac{y_2 - y_1}{x_2 - x_1}$$

$$m = \frac{2.5 - 2.5}{2 - 5}$$

$m = \frac{0}{-3}$, which equals 0!

The slope of a horizontal line is zero. Why is this true? Since a horizontal line never moves up or down, its y-coordinate will never change. This means the "change in y-coordinates" is always 0. When we divide by the change in x-coordinates to find the slope, our final answer will always be 0.

Vertical Lines

A vertical line goes straight up and down.

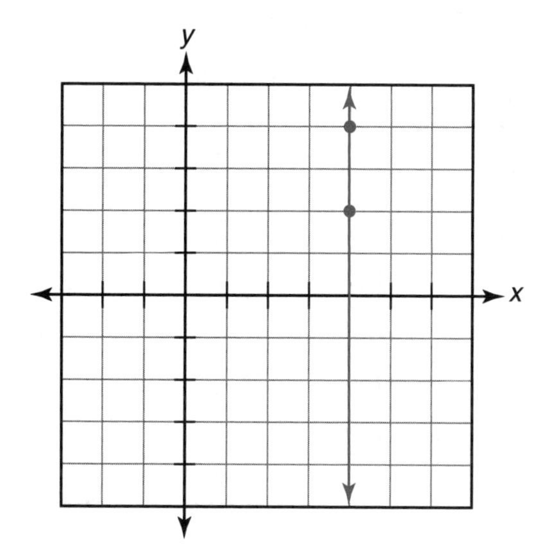

Let's use 2 points on the line to find the slope of the vertical line $x = 4$.

Use (4,2) and (4,4).

$$m = \frac{y_2 - y_1}{x_2 - x_1}$$

$$m = \frac{2 - 4}{4 - 4}$$

$$m = \frac{-2}{0}$$

We can't divide by zero. Any number divided by zero is undefined; therefore, because we cannot divide by zero, **the slope of a vertical line is undefined**. Whenever you try to find the slope of a vertical line, you will get a number divided by zero. Therefore, the slope of every vertical line is undefined!

BRAIN TICKLERS

Set # 53

Find the slope of each line.

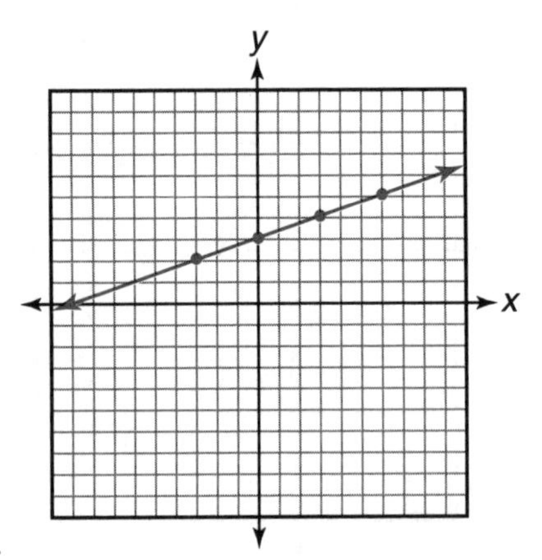

1.

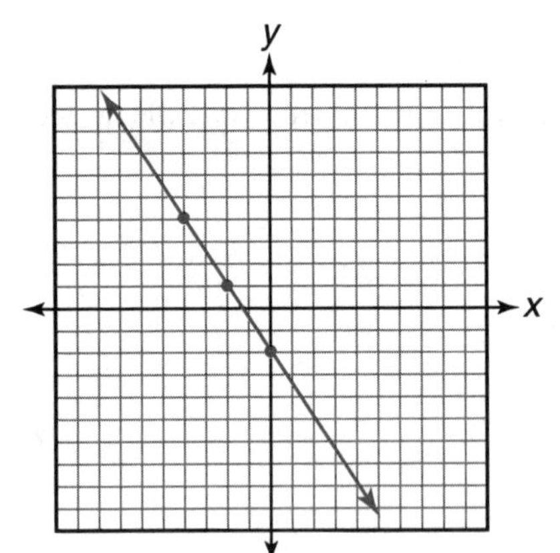

2.

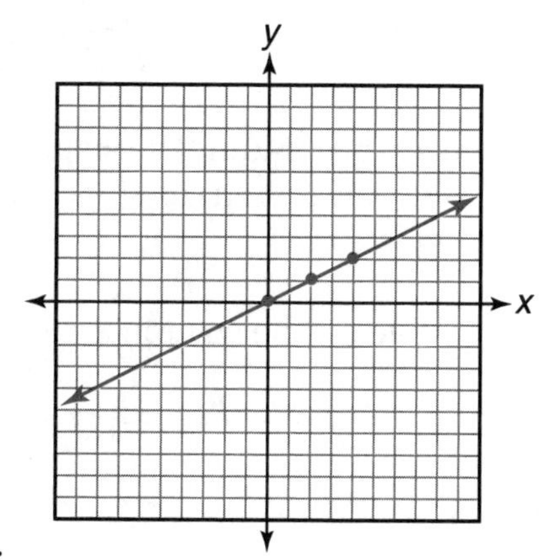

3.

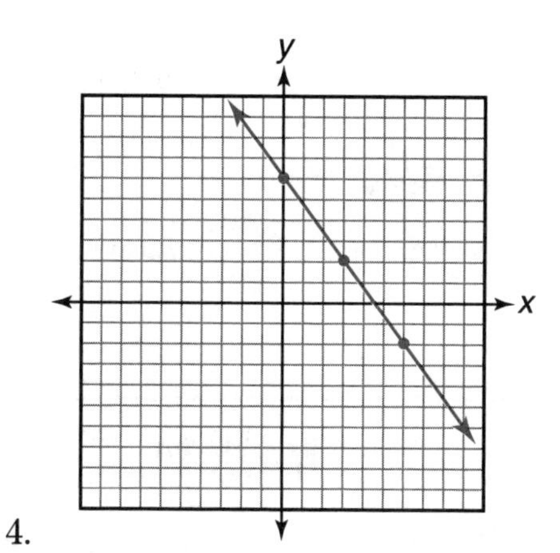

4.

Find the slope of the line passing through the given points.

5. (5,3) and (8,2)

6. (–4,–4) and (5,8)

7. (–3,8) and (–2,6)

8. (8,8) and (7,8)

(Answers are on page 222.)

SLOPE-INTERCEPT FORM OF A LINE

Now that you are a pro at finding the slope of a line using points on a graph, let's see how to find the slope and y-intercept of a line by looking at its equation.

The equation of a line can be written in the form $y = mx + b$.

$$y = mx + b$$

The number in front of x is the slope (m).

b (including the sign) is the y-intercept

When given the equation, you can find the slope and y-intercept.

$y = 2x + 5$

Slope = 2 y-intercept = 5

$y = -3x - 2$

Slope = –3 y-intercept = –2

MATH TALK!

Remember that slope = $\frac{\text{rise}}{\text{run}}$.

A slope of 2 is the same as a slope of $\frac{2}{1}$.

Writing slope as a fraction is helpful when graphing.

If an equation is not written in slope-intercept form, solve for y and rewrite it in that form.

Example

Find the slope and y-intercept of the equation $2x + 4y = 8$. This equation is not in slope-intercept form. Let's change that!

$$2x + 4y = 8$$

Subtract $2x$ from both sides. $\quad -2x \quad -2x$

Divide both sides by 4. $\quad \frac{4y}{4} = \frac{-2x}{4} + \frac{8}{4}$

Simplify. $\quad y = -\frac{1}{2}x + 2$

The equation written in slope-intercept form is $y = -\frac{1}{2}x + 2$. It is easy to see that the slope is $-\frac{1}{2}$ and the y-intercept is 2.

Practice

Now see if you can find the slope and y-intercept. The answers are listed below. Cover them up and test yourself. No cheating!

1. $y = \frac{3}{4}x - 8$

Slope = ______

y-intercept = ______

2. $y = -x + 1$

Slope = ______

y-intercept = ______

3. $2y = 6x - 10$

Slope = ______

y-intercept = ______

ANSWERS

1. Slope = $\frac{3}{4}$
   y-intercept = -8
2. Slope = -1
   y-intercept = 1

3. Slope = 3
   y-intercept = –5

For question 3, make sure you place the line into slope-intercept form, $y = mx + b$. The original equation $2y = 6x - 10$ then becomes $y = 3x - 5$.

BRAIN TICKLERS
Set # 54

Find the slope and y-intercept of each line. Make sure the line is in the form $y = mx + b$.

1. $y = -3x + 1$
2. $2y = 4x + 8$
3. $y = -4 - x$
4. $x + y = 12$

(Answers are on page 222.)

GRAPHING USING THE SLOPE-INTERCEPT

Now that you are a pro finding the slope and y-intercept of a line, let's graph lines using this information.

MATH TALK!

Equation of a line: $y = mx + b$ where m = slope and b = y-intercept

$$\text{Slope} = \frac{\text{Rise}}{\text{Run}}$$

y-intercept = where the line crosses the y-axis

To graph a line using the slope-intercept method, follow these easy steps.

Step 1: Write the equation in the form $y = mx + b$.

Step 2: State the slope.

Step 3: State the y-intercept.

Step 4: Graph the point of the y-intercept. (This will be on the y-axis).

Step 5: From this point, do $\frac{\text{rise}}{\text{run}}$ to find and graph a second point.

Step 6: Draw a line to connect the two points.

Step 7: Label the line.

Example 1

Graph the line $y = \frac{1}{2}x - 3$ using the slope-intercept method.

Slope $= \frac{1}{2}$

y-intercept $= -3$

Start at (0,3) because you are on the y-axis.

From (0,3), go up 1, right 2, and plot a point.

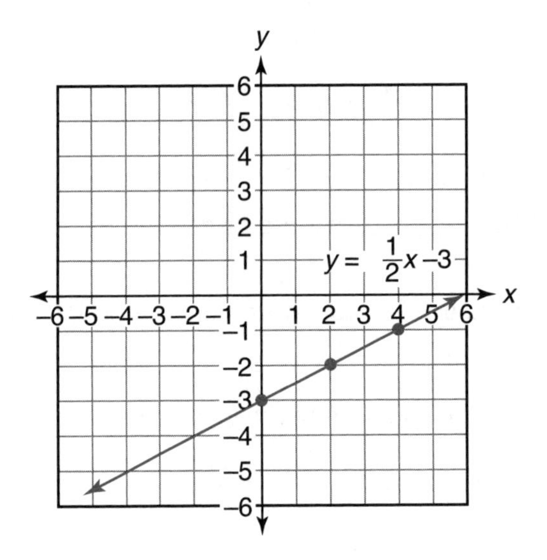

Draw a line connecting the points. Label the line with the equation $y = \frac{1}{2}x - 3$.

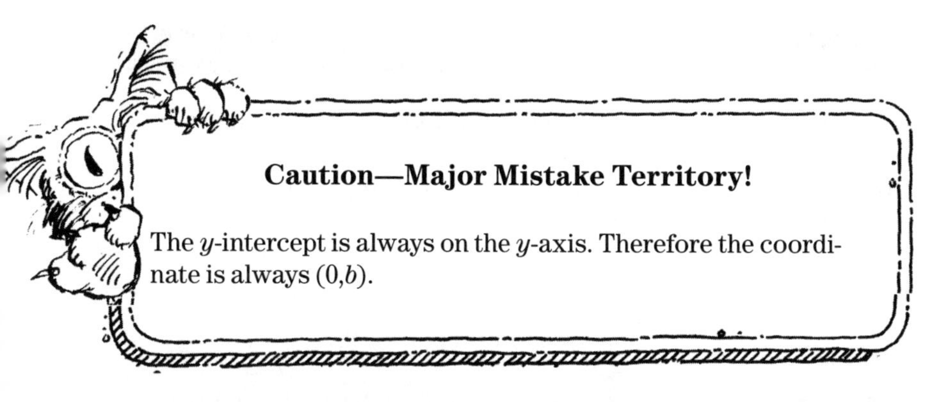

Caution—Major Mistake Territory!

The y-intercept is always on the y-axis. Therefore the coordinate is always $(0,b)$.

Example 2

Graph the line $y = x + 1$ using the slope-intercept method.

Slope = 1

y-intercept = 1

Start at (0,1).

From (0,1), go up 1, right 1, and plot the point.

Draw a line connecting the points. Label the line with the equation $y = x + 1$.

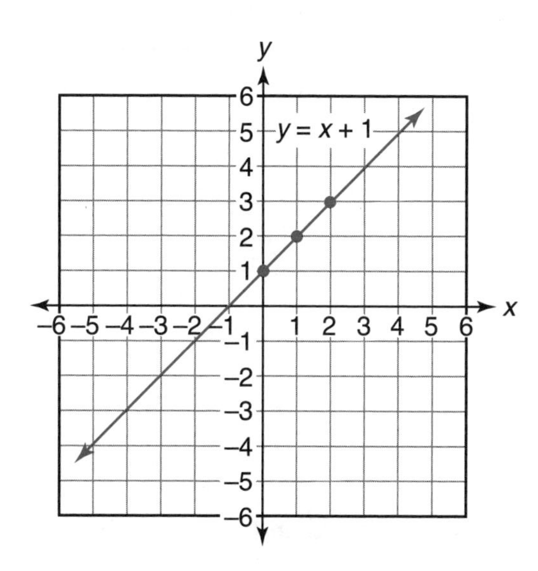

Example 3

Graph the line $2x + 4y = 12$ using the slope-intercept method.

Remember, the line must be written in $y = mx + b$ form. Is this line in the correct form? No!

Rewrite in $y = mx + b$ form.

$$\begin{array}{l} 2x + 4y = 12 \\ \underline{-2x \qquad -2x} \\ \qquad \dfrac{4y}{4} = \dfrac{-2x}{4} + \dfrac{12}{4} \end{array}$$

$$y = -\frac{1}{2}x + 3$$

$$\text{Slope} = -\frac{1}{2}$$

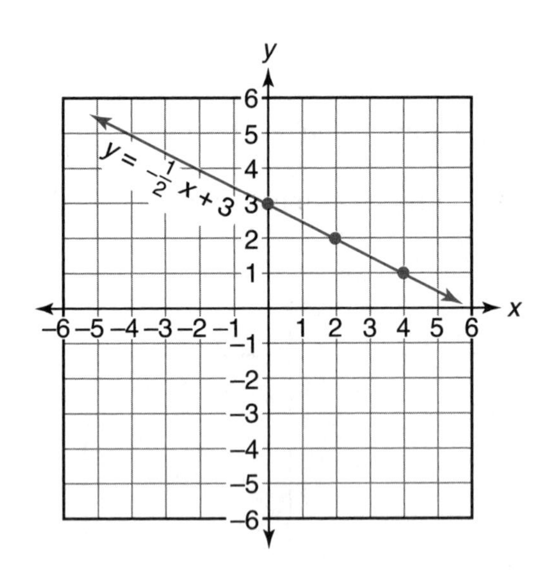

y-intercept = 3

Start at (0,3).

From (0,3), go down 1, right 2, and plot the point.

Draw a line connecting the points. Label the line with the equation $y = -\frac{1}{2}x + 3$.

BRAIN TICKLERS
Set # 55

Graph each line using the slope-intercept method.

1. $y = -3x - 1$
2. $y = \frac{1}{2}x + 2$
3. $x + y = -5$
4. $2x + 2y = 10$

(Answers are on page 223.)

WRITING THE EQUATION OF A LINE

Given the equation of a line, you can find the slope and y-intercept. If you were given the slope of a line and the y-intercept, do you think you could write the equation of the line?

Be a detective.

You were given $y = 2x + 4$.

You found slope = 2 and y-intercept = 4.

What if slope = 3 and y-intercept = 5? What would the equation be?

Think about what you know: $y = mx + b$. Plug it in, plug it in!

The answer is $y = 3x + 5$.

Caution—Major Mistake Territory

For a problem to be an equation, it must include an equals sign.

Exciting Examples

Write the equation of the line.

Slope = –2, y-intercept = 5 → $y = -2x + 5$

Slope = $-\frac{1}{3}x$, y-intercept = –2 → $y = -\frac{1}{3}x - 2$

Slope = 4, y-intercept = 0 → $y = 4x$

Slope = 0, y-intercept = 7 → $y = 7$

BRAIN TICKLERS
Set # 56

Write the equation of each line, given the slope and y-intercept.

1. Slope = 4 y-intercept = 10
2. Slope = $\frac{1}{2}$ y-intercept = 0
3. Slope = –5 y-intercept = –3
4. Slope = $-\frac{2}{3}$ y-intercept = 4
5. Slope = 0 y-intercept = 5

(Answers are on page 225.)

SYSTEMS OF EQUATIONS

A **system of linear equations** consists of two or more lines. The solution to the system is the point(s) that the lines have in common. Three situations are possible.

The lines intersect at one point.

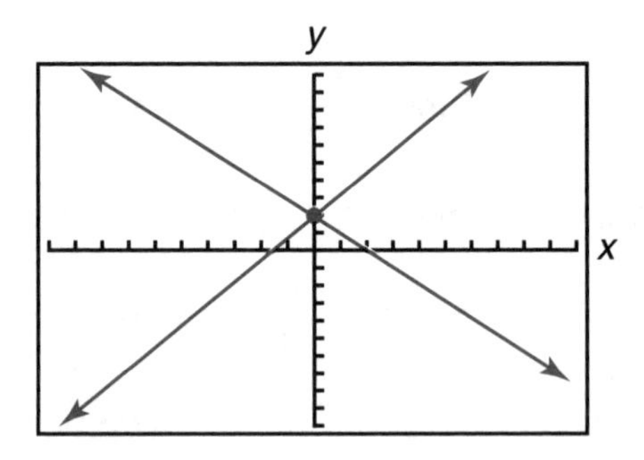

The lines are parallel and never intersect.

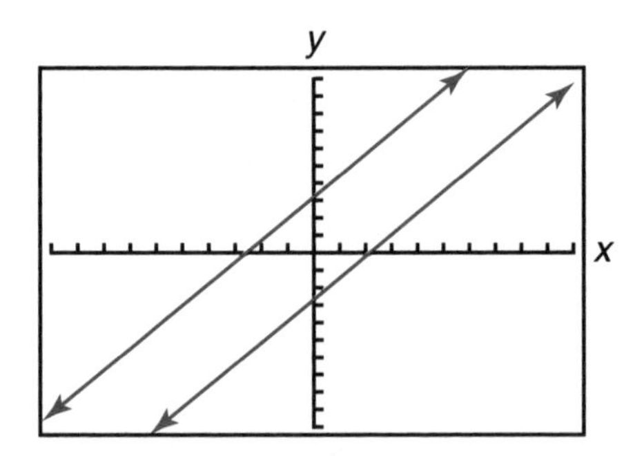

The lines are the same and intersect at every point.

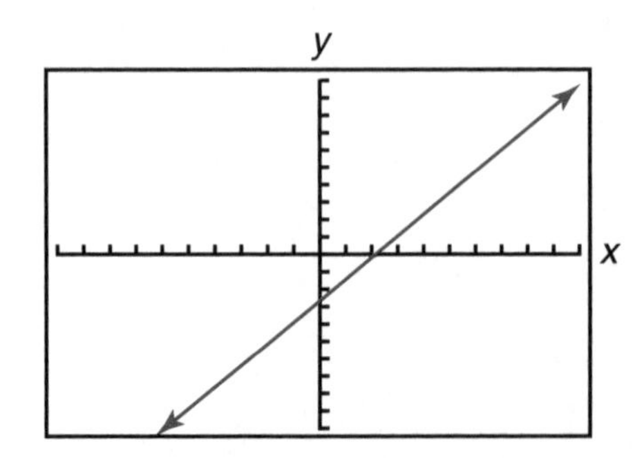

To solve a system of equations:

Step 1: Graph the first equation.

Step 2: Graph the second equation on the same axes.

Step 3: Find the point where they intersect (solution).

Step 4: Check the point in both original equations.

Example 1

Solve graphically $y = 2x + 1$ and $y = -\frac{1}{2}x + 6$.

Step 1: Graph the first line.

$y = 2x + 1 \quad \rightarrow \quad$ Slope = 2, y-intercept = 1

Step 2: Graph the second line.

$y = -\frac{1}{2}x + 6 \quad \rightarrow \quad$ Slope = $-\frac{1}{2}$, y-intercept = 6

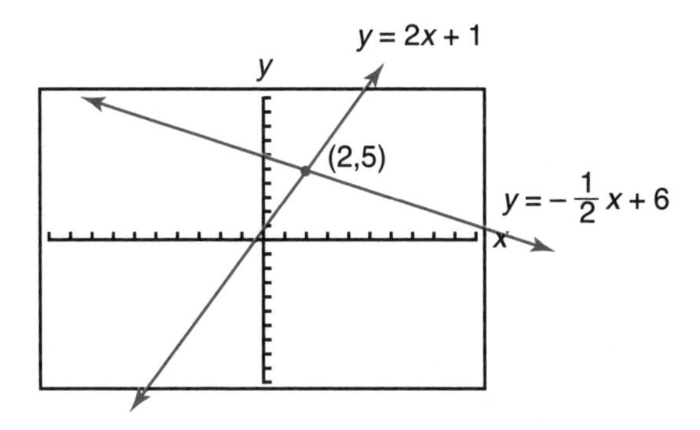

Step 3: Find the point where the graphs intersect. The solution for this system is (2,5).

Step 4: Substitute (2,5) into both equations to check.

$y = 2x + 1$	$y = -\frac{1}{2}x + 6$
$5 = 2(2) + 1$	$5 = -\frac{1}{2}(2) + 6$
$5 = 5$ Check!	$5 = -1 + 6$
	$5 = 5$ Check!

Example 2

Let's try one more.

$$\begin{cases} y = -x + 4 \\ y = 2x + 1 \end{cases}$$

Graph the lines.

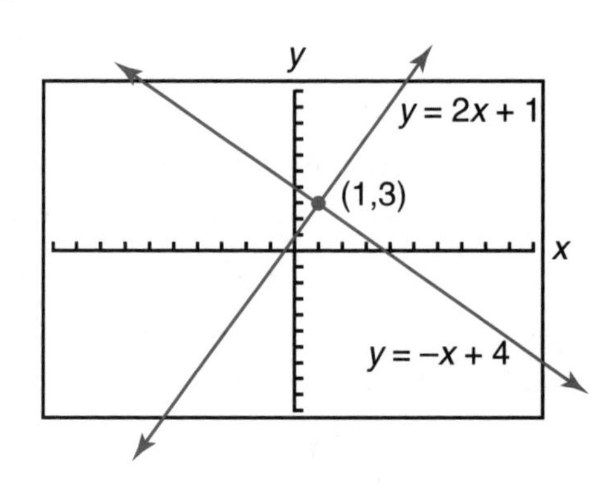

$y = -x + 4$ → Slope = −1, y-intercept = 4

$y = 2x + 1$ → Slope = 2, y-intercept = 1

Find the point of intersection, (1,3).

Check:

$y = -x + 4$	$y = 2x + 1$
$3 = -1 + 4$	$3 = 2(1) + 1$
$3 = 3$ Check!	$3 = 2 + 1$
	$3 = 3$ Check!

BRAIN TICKLERS
Set # 57

Solve graphically and check.

1. $\begin{cases} y = -x + 5 \\ y = 2x - 1 \end{cases}$

2. $\begin{cases} y = 3x + 1 \\ x + y = 1 \end{cases}$

3. $\begin{cases} y = -x + 3 \\ y - 1 = x \end{cases}$

4. Graph the lines $y = 2x + 4$ and $y = 2x - 1$ on the same axes. Explain why there is no solution.

(Answers are on page 225.)

BRAIN TICKLERS—THE ANSWERS

Set # 50, page 183

1. $y = 2x + 1$

x	$2x + 1$	y
–2	–4 + 1	–3
0	0 + 1	1
2	4 + 1	5

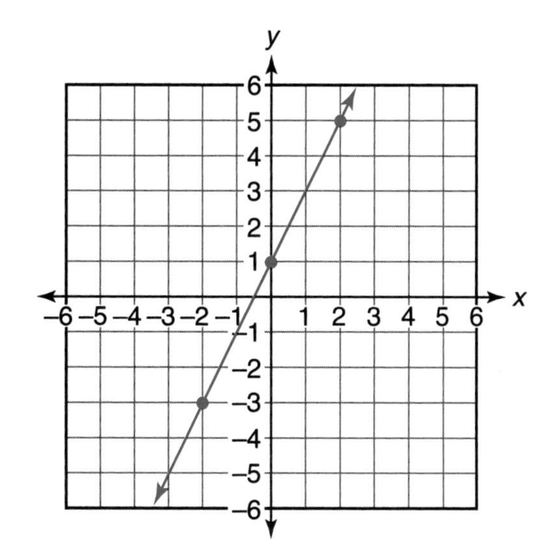

2. $x + y = -2$
   $y = -x - 2$

x	$-x - 2$	y
–2	2 – 2	0
0	0 – 2	–2
2	–2 – 2	–4

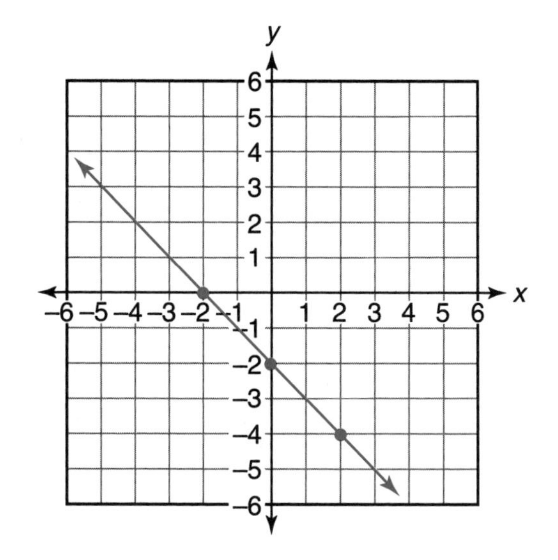

3. $y - 5 = x$
   $y = x + 5$

x	$x + 5$	y
–2	–2 + 5	3
0	0 + 5	5
2	2 + 5	7

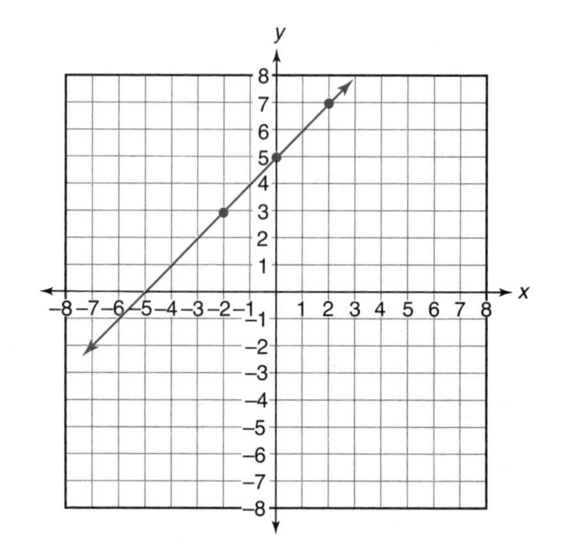

4. $2x - y = 3$
   $y = 2x - 3$

x	$2x - 3$	y
–2	–4 – 3	–7
0	0 – 3	–3
2	4 – 3	1

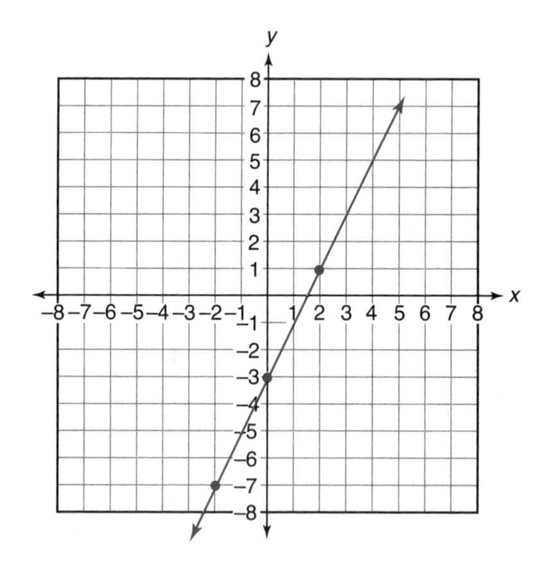

Set # 51, page 187

1. $x = 4$

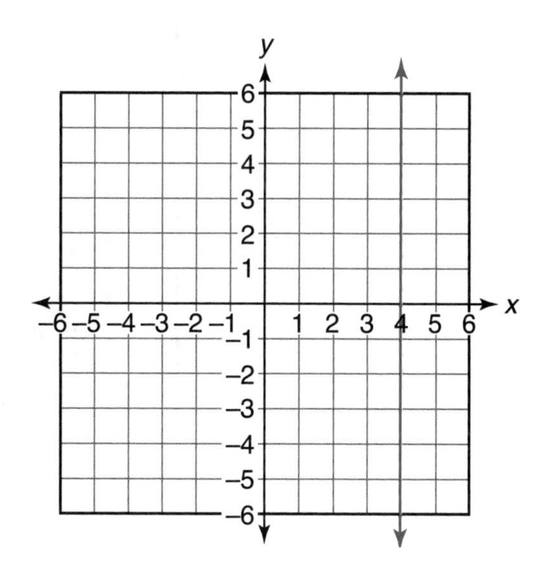

2. $x = -2$

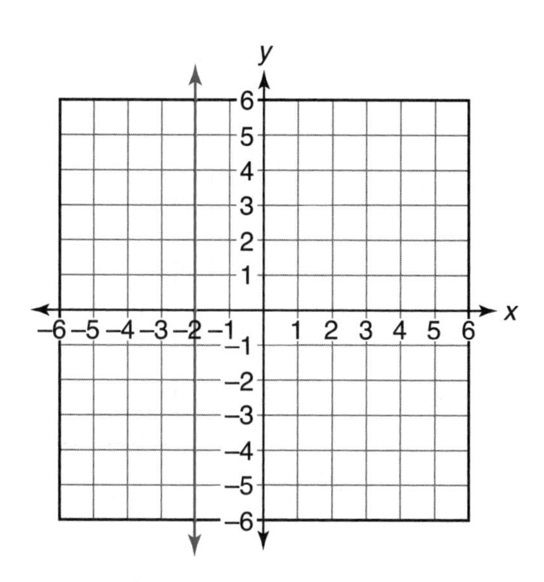

3. $y = 5$

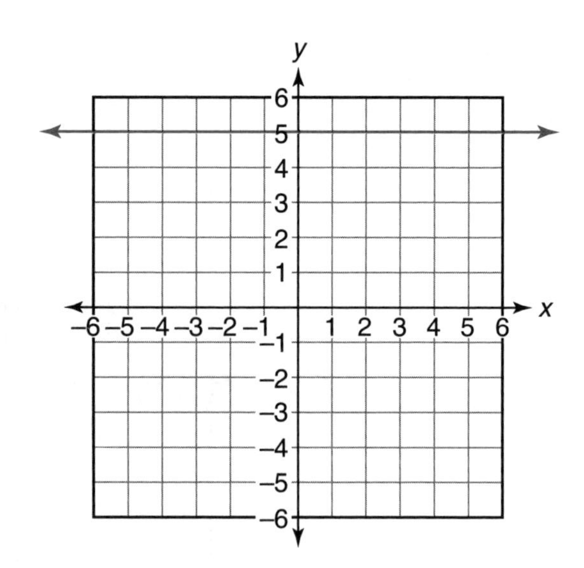

4. $y = -1$

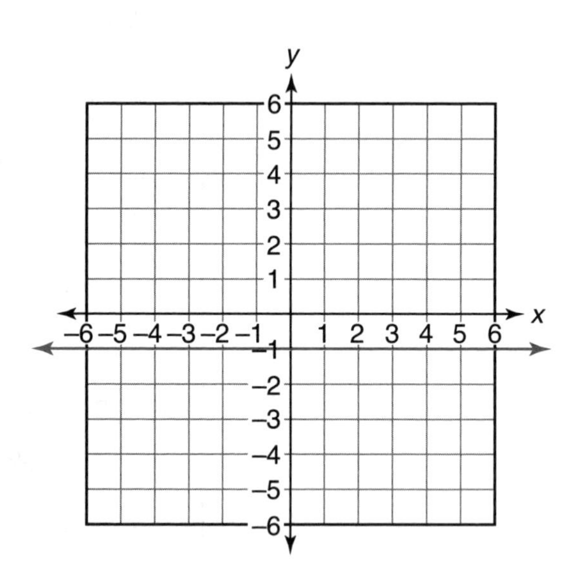

Set # 52, page 190

1. x-intercept = 10
   y-intercept = 5
2. x-intercept = 5
   y-intercept = −5

3. x-intercept = –4
   y-intercept = 8

4. x-intercept = 6
   coordinates of x-intercept = (6,0)
   y-intercept = 3
   coordinates of y-intercept = (0,3)

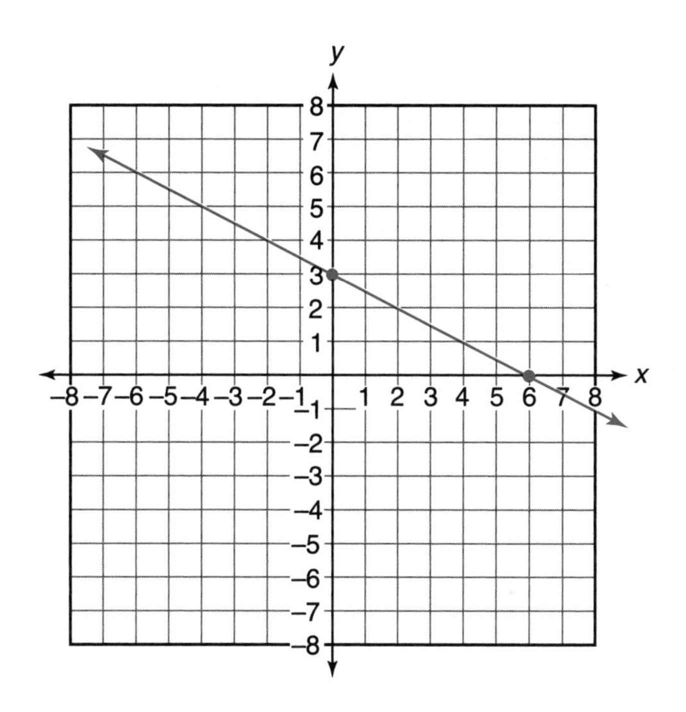

5. x-intercept = –3
   coordinates of x-intercept = (–3,0)
   y-intercept = 3
   coordinates of y-intercept = (0,3)

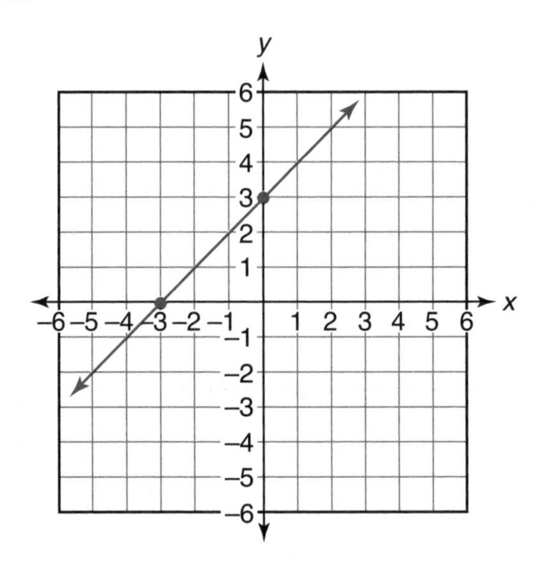

Set # 53, page 202

1. $\frac{1}{3}$
2. $-\frac{3}{2}$
3. $\frac{1}{2}$
4. $-\frac{4}{3}$
5. $-\frac{1}{3}$
6. $\frac{4}{3}$
7. –2
8. 0

Set # 54, page 206

1. Slope = –3, y-intercept = 1
2. Slope = 2, y-intercept = 4
3. Slope = –1, y-intercept = –4
4. Slope = –1, y-intercept = 12

Set # 55, page 211

1. Slope = -3
   y-intercept = -1

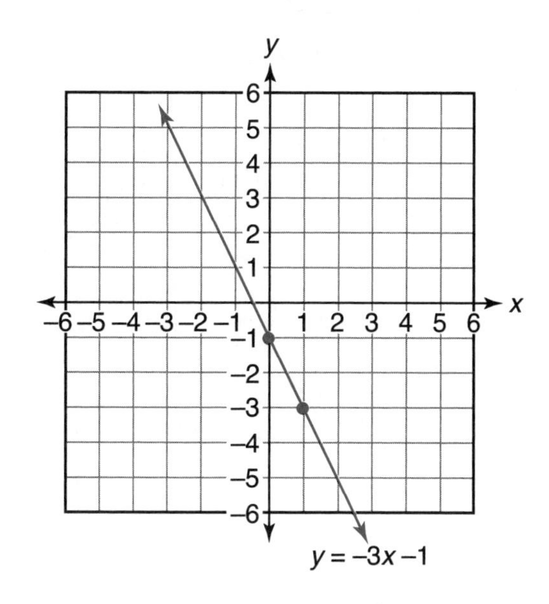

2. Slope = $\frac{1}{2}$
   y-intercept = 2

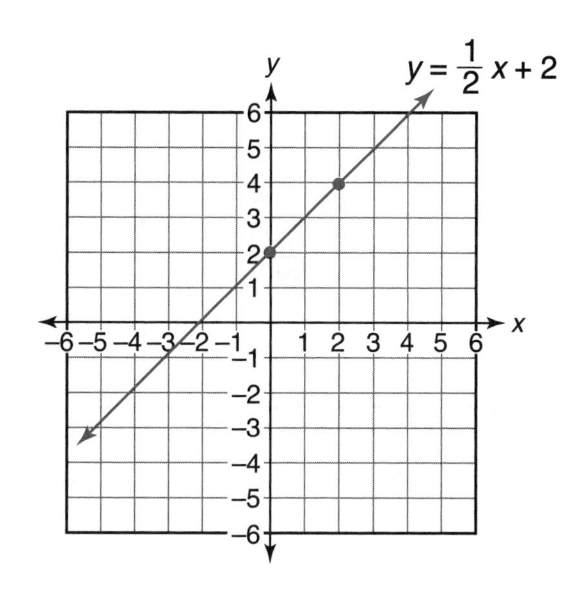

3. Slope = –1
   y-intercept = –5

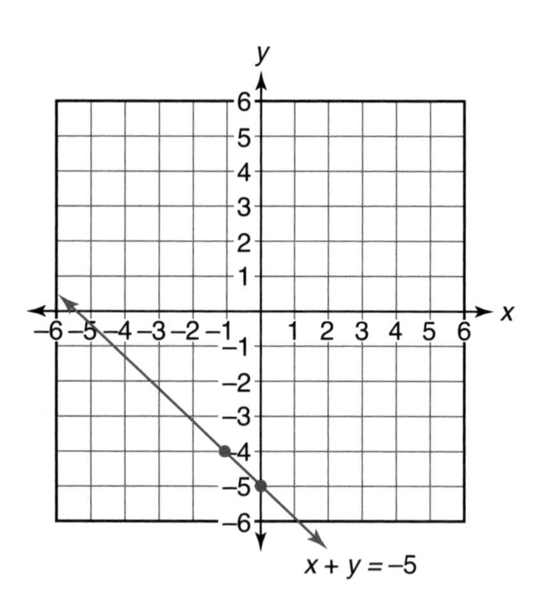

4. Slope = –1
   y-intercept = 5

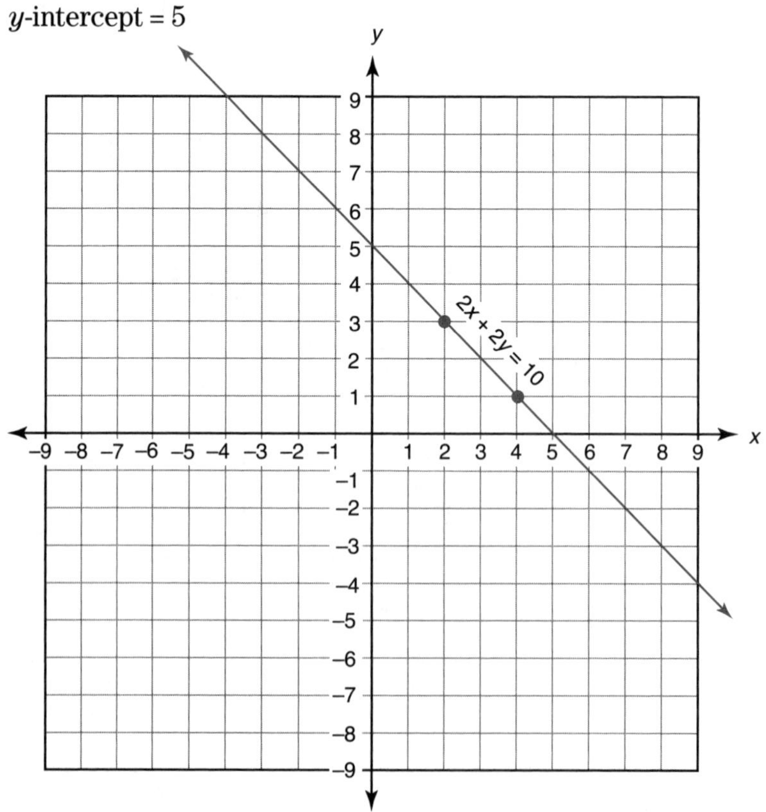

Set # 56, page 213

1. $y = 4x + 10$
2. $y = \frac{1}{2}x$
3. $y = -5x - 3$
4. $y = -\frac{2}{3}x + 4$
5. $y = 5$

Set # 57, page 216

1. (2,3)

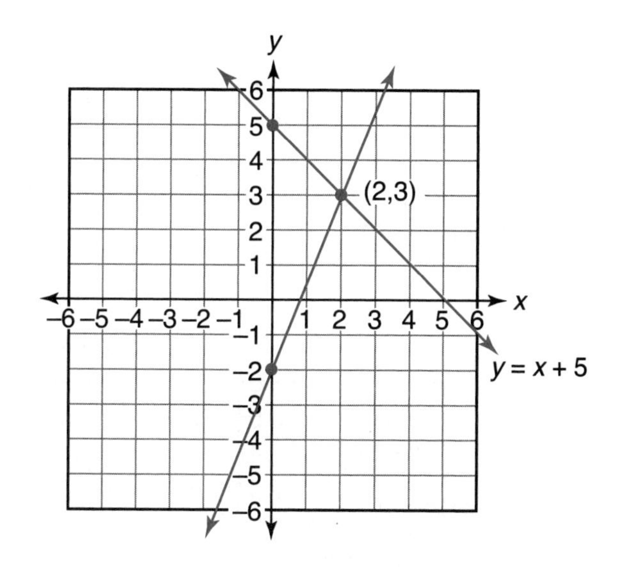

2. (0,1)

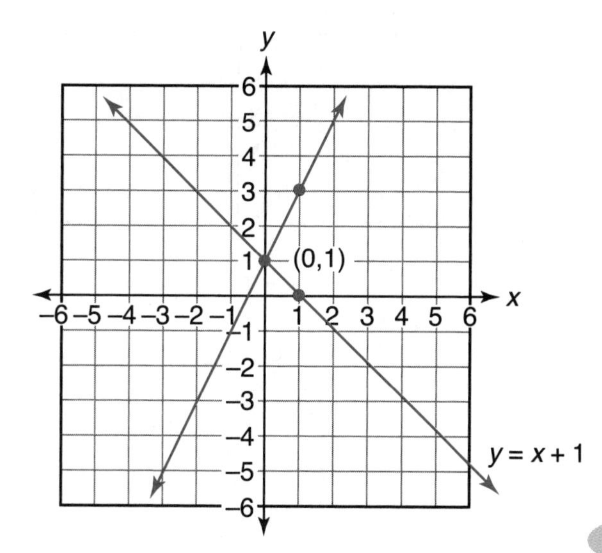

3. (1,2)

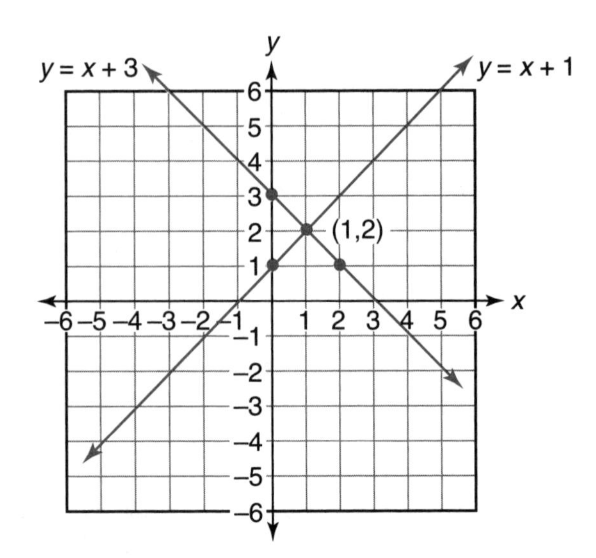

4. The lines are parallel. Since they never interact, they have no solution.

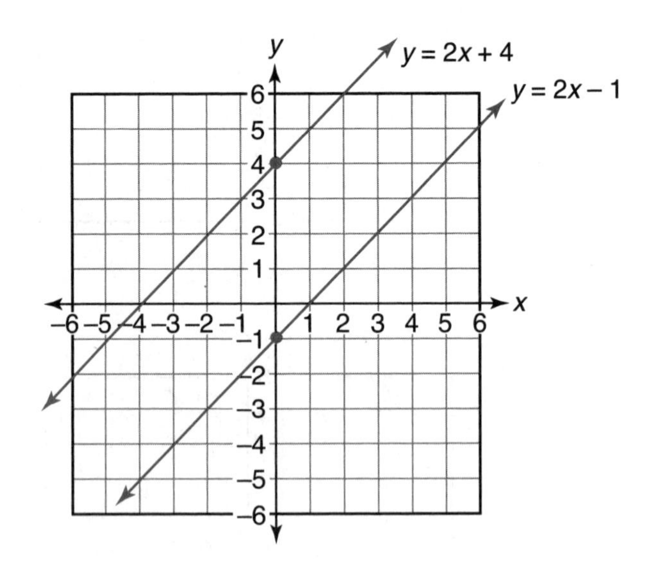

CHAPTER NINE
Polynomials
DRAT, FOIL'ED AGAIN!
(x+4)(x+5) = x² + 9x + 20

POLYNOMIALS

Examples of expressions are $2x^2$ or $6xy$. A polynomial is created when you combine these expressions by addition or subtraction. Each piece or part of the polynomial is called a **term**. Here is an example of a polynomial:

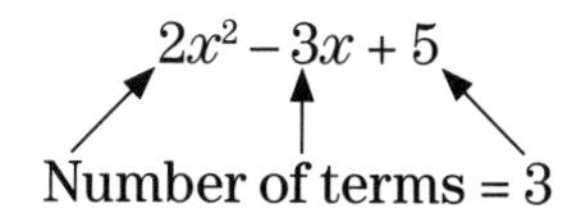

A polynomial can be named according to how many terms it contains.

MONOMIAL

A monomial has one term. It is a number, a variable, or the product of a number and one or more variables.

Examples: x^2 $\quad 4a^3b$ $\quad \frac{1}{3}xy$ $\quad -0.25$ $\quad 15$

Notice there are no negative exponents, no fractional exponents, and no variables in the denominator.

BINOMIAL

Think bicycle, with two wheels! A binomial contains two unlike terms connected by addition or subtraction.

Examples: $3x + 1$ $\quad x^2 - 2x$ $\quad 4x + 3y$ $\quad 5 + \frac{1}{2}z$

TRINOMIAL

Think tricycle, with three wheels! A trinomial has three unlike terms connected by addition or subtraction or both.

Examples: $x^2 + 5x + 4$ $\quad 2x + 5y - 7$ $\quad a + b + 2c$

POLYNOMIAL

A polynomial is a monomial or the sum of monomials whose exponents are positive. Polynomials can have one or more terms.

CLASSIFYING POLYNOMIALS

POLYNOMIAL	CLASSIFIED BY NUMBER OF TERMS
6	Monomial
$-2x$	Monomial
$3x + 1$	Binomial
$x^2 + 2x - 5$	Trinomial
$4x^3 - 8x$	Binomial
$2x^4 - 7x^3 - 5x + 1$	Polynomial

BRAIN TICKLERS
Set # 58

Classify each polynomial by its number of terms.

1. $2x + 4$
2. $x^2 + 3x + 5$
3. $a + 2b - 4c + 12$
4. $-abc$

(Answers are on page 249.)

ADDING AND SUBTRACTING POLYNOMIALS

If you are going to add or subtract monomials or polynomials, you need to have **like terms**.

LIKE TERMS	UNLIKE TERMS
$2xy$, $4xy$, $-3xy$ → all have xy	x^2, x, $2x^3$ → all x's have different exponents
$4a^2b^5$, $10a^2b^5$ → all have a^2b^5	$4ab^2$, $3a^2b$, $10a^5b^5$ → the variables have different exponents

To add or subtract polynomials, you have to combine all like terms. A good strategy to use is to circle, box, or underline like terms.

Example 1

Simplify: $3xy + 5y - 2xy + 11y$

$\bigcirc 3xy \; \boxed{+ 5y} \; \bigcirc - 2xy \; \boxed{+ 11y}$

Find the like terms: Simplify $3xy - 2xy = xy$
Simplify $5y + 11y = 16y$

The answer is $xy + 16y$.

Example 2

Add: $(4x^2 + 6x + 9) + (2x^2 + 5x + 8)$

Add $4x^2 + 2x^2 = 6x^2$
Add $6x + 5x = 11x$
Add $9 + 8 = 17$

The answer is $6x^2 + 11x + 17$.

A HELPFUL HINT

Instead of circling and boxing like terms, stack them!

Problem: $(2x^2 + 4x + 7) + (3x^2 - 8x + 10)$

Stack them according to like terms!

$$\begin{array}{r} 2x^2 + 4x + 7 \\ \underline{3x^2 - 8x + 10} \\ 5x^2 - 4x + 17 \end{array}$$

Now add each column! $5x^2 - 4x + 17$

Subtracting polynomials is exactly the same as adding; you have to combine like terms. You have to be careful because you are subtracting everything in the second parentheses. So the minus sign has to be distributed to each term in parentheses.

Let's review distributing!

$-2(-3x - 5) \rightarrow$ Multiply $-2(-3x) = 6x$
Multiply $-2\,(-5) = 10$
Answer $= 6x + 10$

Example 3

Subtract: $(3x^2 + 4x - 5) - (2x^2 - 6x + 3)$ Distribute the minus sign to each term in the second parentheses.

$$3x^2 + 4x - 5 - 2x^2 + 6x - 3$$

Now combine like terms:
$3x^2 - 2x^2 = x^2$
$4x + 6x = 10x$
$-5 - 3 = -8$

The answer is $5x^2 + 10x - 8$.

Stacking also works with subtraction. Just remember to change the signs in the second parentheses.

Example 4

$(4x^4 + 5x - 8) - (-2x^2 + 6x - 9)$

$$\begin{array}{l} 4x^4 + 5x - 8 \\ \underline{2x^2 - 6x + 9} \\ 6x^2 - x + 1 \end{array}$$

Caution—Major Mistake Territory!

When subtracting, you have to remember to change the signs first!

$$(14x - 5) - (7x + 8)$$
$$14x - 5 - 7x - 8$$
$$7x - 13$$

BRAIN TICKLERS

Set # 59

Add or subtract.

1. $(3x^2 + 7x - 10) + (4x^2 - 10x - 14)$
2. $(-5x^2 - 8x + 4) + (7x^2 + 4x - 12)$
3. $(7x^2 + 10x + 5) - (4x^2 + 8x - 6)$
4. $(6x^2 - 2x + 1) - (x^2 - 3x + 10)$

5. Add:

$$\begin{array}{r} 4x^2 + 8x - 9 \\ \underline{3x^2 - 6x + 10} \end{array}$$

6. Subtract:

$$\begin{array}{r} 5x^2 - 7x + 10 \\ \underline{2x^2 - 8x - 12} \end{array}$$

(Answers are on page 249.)

MULTIPLYING POLYNOMIALS BY A MONOMIAL

In our exponent unit, we multiplied monomials by monomials.

Multiply: $(2x^2)(4x^5)$ $= 8x^7$ Multiply the coefficients, and add the exponents!

Now we want to multiply polynomials by a monomial. This chart will show the different types of problems you may see.

MULTIPLYING POLYNOMIALS BY MONOMIALS

Type	Example	How to Solve
(Monomial)(Monomial)	$(5x^2)(6x^4) =$ $30x^6$	Multiply coefficients, add exponents
(Monomial)(Binomial)	$x(2x + 3) =$ $2x^2 + 3x$	Distribute! $x(2x) = 2x^2$ $x(3) = 3x$
(Monomial)(Trinomial)	$3x(2x^2 + 4x - 5) =$ $6x^3 + 12x^2 - 15x$	Distribute to each term $3x(2x^2) = 6x^3$ $3x(4x) = 12x^2$ $3x(-5) = -15x$
(Monomial)(Polynomial)	$2x^2(x^4 - 3x^3 + 2x^2 - 4) =$ $2x^6 - 6x^5 + 4x^4 - 8x^2$	Distribute $2x^2$ to each term

Exciting Examples

Let's try a few.

Multiply: $2x(x - 4)$

Multiply $2x$ by each term inside the parentheses.

$2x(x) = 2x^2$ and $2x(-4) = -8x$, so the answer is $2x^2 - 8x$.

Multiply: $-3x^2(4x + 3y + 10)$

Multiply $-3x^2$ by each term inside the parentheses.

$-3x^2(4x) = -12x^3$ and $-3x^2(3y) = -9x^2y$ and $-3x^2(10) = -30x^2$. So the answer is $-12x^3 - 9x^2y - 30x^2$.

BRAIN TICKLERS
Set # 60

1. $(2x^3y^2)(6x^5y^3)$
2. $(-9a^4b^3)(2a^7b^4)$
3. $3m(5m^2 + 4)$
4. $3n(2n^4 - 7n^2)$
5. $-5y(y^2 + 3y - 8)$
6. $2x^2(-4x^4 + 3x^3 - 5x^2 + 6x - 12)$

(Answers are on page 250.)

MULTIPLYING BINOMIALS

Follow these easy steps to multiply binomials.

Step 1: Multiply the First two terms. F → first

Step 2: Multiply the Outside two terms. O → outside

Step 3: Multiply the Inside two terms. I → inside

Step 4: Multiply the Last two terms. L → last

Step 5: Simplify by adding the like terms.

Example 1

Let's try it!

Multiply: $(x + 3)(x + 2)$

First: $(x + 3)(x + 2)$ Multiply $(x)(x) = x^2$

Outside: $(x + 3)(x + 2)$ Multiply $(2)(x) = 2x$

Inside: $(x + 3)(x + 2)$ Multiply $(3)(x) = 3x$

Last: $(x + 3)(x + 2)$ Multiply $(3)(2) = 6$

Combine like terms by adding: $x^2 + 5x + 6$

See how easy FOIL is? First, Outside, Inside, Last!

Example 2

Multiply:	$(x-3)(x+4)$	
First:	$(x-3)(x+4)$	Multiply $(x)(x) = x^2$
Outside:	$(x-3)(x+4)$	Multiply $(4)(x) = 4x$
Inside:	$(x-3)(x+4)$	Multiply $(-3)(x) = -3x$
Last:	$(x-3)(x+4)$	Multiply $(-3)(4) = -12$

Combine like terms by adding: $x^2 + x - 12$

Caution—Major Mistake Territory!

When a problem has subtraction in it, like $x - 3$, remember to include the sign in front of the number when multiplying. So think of minus 3 as -3.

Example 3

Multiply:	$(x-5)(x-4)$	
First:	$(x-5)(x-4)$	Multiply $(x)(x) = x^2$
Outside:	$(x-5)(x-4)$	Multiply $(-4)(x) = -4x$
Inside:	$(x-5)(x-4)$	Multiply $(-5)(x) = -5x$
Last:	$(x-5)(x-4)$	Multiply $(-5)(-4) = 20$

Combine like terms by adding: $x^2 - 9x + 20$

Example 4

What if the problem was written as $(x - 3)^2$? Remember squared means to multiply the problem by itself. So $(x - 3)^2$ is the same as $(x - 3)(x - 3)$. Write it out twice so you can use FOIL!

First: $(x - 3)(x - 3)$ Multiply $(x)(x) = x^2$

Outside: $(x - 3)(x - 3)$ Multiply $(-3)(x) = -3x$

Inside: $(x - 3)(x - 3)$ Multiply $(-3)(x) = -3x$

Last: $(x - 3)(x - 3)$ Multiply $(-3)(-3) = 9$

Combine like terms by adding: $x^2 - 6x + 12$

BRAIN TICKLERS

Set # 61

Multiply.

1. $(x + 3)(x + 4)$
2. $(x - 2)(x + 5)$
3. $(x - 4)(x - 5)$
4. $(x - 7)(x + 4)$
5. $(x + 5)(x - 5)$
6. $(x + 3)^2$

(Answers are on page 250.)

DIVIDING A POLYNOMIAL BY A MONOMIAL

Let's review how to divide a monomial by a monomial.

Divide: $\frac{20x^5y^4}{5x^2y^3}$ Divide the coefficients, and subtract the exponents!

$\frac{20}{5} = 4$ $\frac{x^5}{x^2} = x^{5-2} = x^3$ $\frac{y^4}{y^3} = y^{4-3} = y$

Answer: $4x^3y$

To divide a polynomial by a monomial, you have to use these same rules. Divide the coefficients, and subtract the exponents.

Example 1

Divide: $\frac{12x^3 + 6x^2 + 3x}{3x}$

If the entire fraction is divided by $3x$, then each term is divided by $3x$. You can rewrite the problem as 3 fractions.

Divide: $\frac{12x^3 + 6x^2 + 3x}{3x}$

$$\frac{12x^3}{3x} + \frac{6x^2}{3x} + \frac{3x}{3x}$$

Now divide the numbers and subtract the exponents!

$\frac{12x^3}{3x} = 4x^2$ $\frac{6x^2}{3x} = 2x$ $\frac{3x}{3x} = 1$

Answer: $4x^2 + 2x + 1$

Use these steps to help you divide polynomials by monomials:

Step 1: Rewrite the problems as separate fractions.

Step 2: Divide each term in the numerator by the monomial in the denominator.

Step 3: Simplify each fraction.

Example 2

Divide: $\dfrac{10x^4y + 12x^5y^2 - 8xy^2}{2xy}$

Rewrite as separate fractions: $\dfrac{10x^4y}{2xy} + \dfrac{12x^5y^2}{2xy} - \dfrac{8xy^2}{2xy}$

Divide the coefficients, and subtract the exponents:
$5x^3 + 6x^4y - 4y$

Example 3

Divide: $\dfrac{18x^6 - 24x^4 + 12x^3 + 6x^2}{6x}$

Rewrite as separate fractions: $\dfrac{18x^6}{6x} - \dfrac{24x^4}{6x} + \dfrac{12x^3}{6x} + \dfrac{6x^2}{6x}$

Divide the coefficients, and subtract the exponents:
$3x^5 - 4x^3 + 2x^2 + 6x$

BRAIN TICKLERS
Set # 62

Divide.

1. $\dfrac{24x^5 + 12x^2 - 4x}{4}$
2. $\dfrac{15x^4y^2 - 25x^3y + 5xy}{5xy}$
3. $\dfrac{4a^3 - 2a + 12}{2}$

(Answers are on page 250.)

GCF

What are the factors of 12? 1, 2, 3, 4, 6, and 12. The factors of 12 are the numbers, when multiplied, that equal 12. Let's review some important vocabulary.

When two numbers are multiplied together, the answer is called the **product**.

The integers that are multiplied together are called the **factors** of the product.

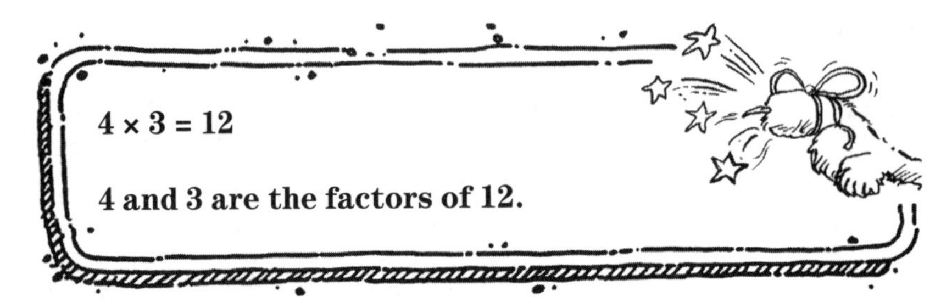

The **greatest common factor (GCF)** of two or more numbers is the largest number that is a factor of both (or all) numbers.

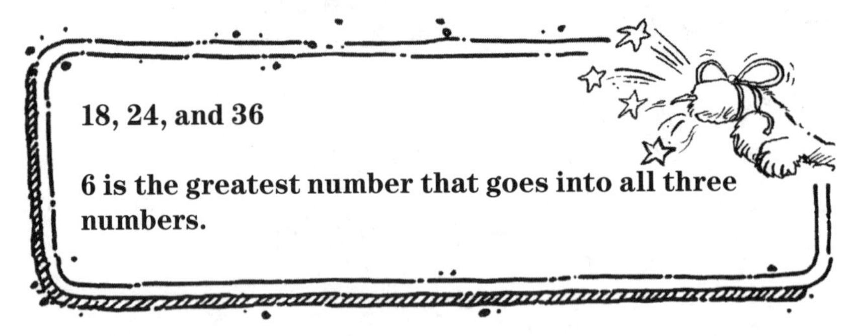

Now, let's add some variables! What is the greatest common factor of $40x^2$ and $28x$?

Factors of $40x^2$ = 1, 2, 4, 5, 8, 10, 20, 40, x, x

Factors of $28x$ = 1, 2, 4, 7, 14, 28, x

The greatest number and variable common to both is 4 and x. Multiply these together to get the GCF, which is $4x$.

Exciting Example

Find the GCF of $15x^3y^2$ and $25x^2y^4$.

Factors of $15x^3y^2 = 1, 3, 5, 15, x, x, x, y, y$

Factors of $25x^2y^4 = 1, 5, 25, x, x, y, y, y, y$

The greatest common factors are $5, x, x, y, y$. So the GCF is $5x^2y^2$

MATH TALK!

For GCF, think of the greatest factor that divides into each number. For the variables, find the most you can take from each term.

$12y^4, 8y^3$

4 is the highest number that divides into 12 and 8.
y^3 is the most you can take from each variable equally.
So $4y^3$ is the GCF.

To factor a polynomial, follow these easy steps:

Step 1: Find the GCF of the polynomial.

Step 2: Divide each term by the GCF (divide the coefficients, subtract the exponents).

Step 3: Write the polynomial as the product of the GCF and the quotient.

Factor $4x^2 + 8x$.

Step 1: Find the GCF of $4x^2$ and $8x$.

$4x^2 = 1, 2, 4, x, x$ / $8x = 1, 2, 4, 8, x$ } The common factors are 4 and x, so the GCF = $4x$.

Step 2: Divide $4x^2 + 8x$ by $4x$.

$$\frac{4x^2 + 8x}{4x}$$

$$\frac{4x^2}{4x} + \frac{8x}{4x}$$

$$x + 2$$

Step 3: Write the polynomial $4x^2 + 8x$ as $4x(x + 2)$.

MATH TALK!

Think of factoring as the reverse of multiplying.

$$4x(x + 2) = 4x^2 + 8x$$

Factor $15xy + 20x^2y$

Step 1: Find the GCF of $15xy$ and $20x^2y$.

$15xy = 1, 3, 5, 15, x, y$ / $20x^2y = 1, 2, 4, 5, 10, 20, x, x, y$ } The common factors are 5, x, and y, so the GCF = $5xy$

Step 2: Divide $15xy + 20x^2y$ by $5xy$.

$$\frac{15xy + 20x^2y}{5xy}$$

$$\frac{15xy}{5xy} + \frac{20x^2y}{5xy}$$

$$3 + 4x$$

Step 3: Write in factored form: $5xy(3 + 4x)$.

BRAIN TICKLERS
Set # 63

1. What is the GCF of $15x$ and $35xy$?

2. What is the GCF of $6x^2$ and $18x^3$?

3. What is the GCF of $10x^3y^2$ and $15xy$

4. What is the GCF of $24x^3y^2z$ and $36x^2y^2z^2$?

Factor each of the following.

5. $2x + 4y$

6. $18x^2y + 27xy$

7. $12x^3 + 16x^5$

8. $14abc + 16a^2b$

(Answers are on page 250.)

FACTORING TRINOMIALS

First you learned how to multiply binomials by using FOIL.

Multiply: $(x + 2)(x + 1) = x^2 + 3x + 2$

Now we are going to factor trinomials. This is basically the opposite of FOIL. The trinomial will factored as the product of two binomials.

(? ?)(? ?)

To factor, a helpful hint is to think it's "factor time!" Think **AM** (like the morning). **A** stands for **add**, and **M** stands for **multiply**. Write A over the 2nd term and M over the 3rd term. This will help us find the 2nd and 3rd term.

Example 1

A M
Factor: $x^2 + 5x + 6$

(? ?)(? ?)

Step 1: The first term of each binomial will be the factors of x^2 $\rightarrow x(x) = x^2$.

$(x \quad ?)(x \quad ?)$

Step 2: Now, find numbers that multiply to 6 and add to 5. We must also consider the signs so that the factors will multiply to positive 6 and add to positive 5.

The factors will be 3 and 2 because $3(2) = 6$ and $3 + 2 = 5$.

$x^2 + 5x + 6 = (x + 3)(x + 2)$

The order of the factors does not matter. $(x + 2)(x + 3)$ will FOIL to give the same answer.

MATH TALK!

The phrase "factor time" helps you remember to use AM over the second and third terms. "A" stands for adding to find the 2nd term. "M" stands for multiplying to find the 3rd term. This works for factoring only trinomials. If you have 3 terms, this is a great strategy to help you remember how to factor.

A M
$x^2 + 5x + 6$

Example 2

Let's try another one. Factor $x^2 - 7x + 12$.

A M
Think trinomial, factor time! $x^2 - 7x + 12$

(? ?)(? ?)

Step 1: The first term of each binomial will be the factors of x^2 $\rightarrow x(x) = x^2$.

$(x \quad ?)(x \quad ?)$

Step 2: Find numbers (and signs) that multiply to 12 and add to –7 (think of positives and negatives).

Factors of 12 → 1, 12 → will not add to –7
6, 2 → will not add to –7
4, 3 → could add to –7 if the signs were correct signs

$-4 + -3 = -7$, so the factors are –4 and –3

$x^2 - 7x + 12 = (x - 4)(x - 3)$

Example 3

Factor $x^2 - 2x - 8$

A M

Trinomial—factor time! $x^2 - 2x - 8$

(? ?)(? ?)

Step 1: What will multiply to give you the first term x^2? $x(x)$
$(x \quad ?)(x \quad ?)$

Step 2: Find numbers that multiply to –8 and add to –2.
8(1) → will not add to –2

4(2) → will multiply to 8.
Now think of the signs. –4(2) = –8 and –4 + 2 = –2

The factors are $(x - 4)(x + 2)$.

Caution—Major Mistake Territory!

When factoring, it is easiest to first figure out the numbers that add to the middle term and multiply to the last term. Once you know you have the correct numbers, then figure out the signs!

A M
$x^2 - 7x + 12$

4 and 3 multiply to 12 and add up to 7.
How would you get –7 and +12?
–4 plus –3 = –7 and –4(–3) = 12, so $(x - 4)(x - 3)$.

Let's look at several examples with different signs.

A M
$x^2 + 6x + 5$
$(x + 1)(x + 5)$

A M
$x^2 - 5x + 6$
$(x - 3)(x - 2)$

A M
$x^2 - 3x - 28$
$(x - 7)(x + 4)$

A M
$x^2 + 2x - 15$
$(x + 5)(x - 3)$

BRAIN TICKLERS
Set # 64

Factor.

1. $x^2 + 13x + 42$
2. $x^2 + 11x + 18$
3. $x^2 - 9x + 20$
4. $x^2 - 11x + 28$
5. $x^2 - 2x - 15$
6. $x^2 - 3x - 28$
7. $x^2 + 3x - 40$
8. $x^2 + 10x - 11$

(Answers are on page 250.)

BRAIN TICKLERS—THE ANSWERS

Set # 58, page 230

1. Binomial
2. Trinomial
3. Polynomial
4. Monomial

Set #59, page 233

1. $7x^2 - 3x - 24$
2. $2x^2 - 4x - 8$
3. $3x^2 + 2x + 11$
4. $5x^2 + x - 9$
5. $7x^2 + 2x + 1$
6. $3x^2 + x + 22$

Set # 60, page 236

1. $12x^8y^5$
2. $-18a^{11}b^7$
3. $15m^3 + 12m$
4. $6n^5 - 21n^3$
5. $-5y^3 - 15y^2 + 40y$
6. $-8x^6 + 6x^5 - 10x^4 + 12x^3 - 24x^2$

Set # 61, page 239

1. $x^2 + 7x + 12$
2. $x^2 + 3x - 10$
3. $x^2 - 9x + 20$
4. $x^2 - 3x - 28$
5. $x^2 - 25$
6. $x^2 + 6x + 9$

Set #62, page 241

1. $6x^5 + 3x^2 - x$
2. $3x^3y - 5x^2 + 1$
3. $2a^3 - a + 6$

Set #63, page 245

1. $5x$
2. $6x^2$
3. $5xy$
4. $12x^2y^2z$
5. $2(x + 2y)$
6. $9xy(2x + 3)$
7. $4x^3(3 + 4x^2)$
8. $2ab(7c + 8a)$

Set #64, page 249

1. $(x + 7)(x + 6)$
2. $(x + 9)(x + 2)$
3. $(x - 5)(x - 4)$
4. $(x - 7)(x - 4)$
5. $(x - 5)(x + 3)$
6. $(x - 7)(x + 4)$
7. $(x + 8)(x - 5)$
8. $(x + 11)(x - 1)$

INDEX

Really. This isn't going to hurt at all . . .

Learning won't hurt when middle school and high school students open any *Painless* title. These books transform subjects into fun—emphasizing a touch of humor and entertaining brain-tickler puzzles that are fun to solve.

Extra bonus—each title followed by (*) comes with a FREE app!
Download a fun-to-play arcade game to your iPhone, iTouch, iPad, or Android™ device. The games reinforce the study material in each book and provide hours of extra fun.

Each book: Paperback

Painless Algebra, 3rd Ed.*
Lynette Long, Ph.D.
ISBN 978-0-7641-4715-9, $9.99, *Can$11.99*

Painless American Government
Jeffrey Strausser
ISBN 978-0-7641-2601-7, $9.99, *Can$11.99*

Painless American History, 2nd Ed.
Curt Lader
ISBN 978-0-7641-4231-4, $9.99, *Can$11.99*

Painless Chemistry*
Loris Chen
ISBN 978-0-7641-4602-2, $9.99, *Can$11.99*

Painless Earth Science*
Edward J. Denecke, Jr.
ISBN 978-0-7641-4601-5, $9.99, *Can$11.99*

Painless English for Speakers of Other Languages, 2nd Ed.*
Jeffrey Strausser and José Paniza
ISBN 978-1-4380-0002-2, $9.99, *Can$11.50*

Painless Fractions, 3rd Ed.*
Alyece Cummings, M.A.
ISBN 978-1-4380-0000-8, $9.99, *Can$11.50*

Painless French, 2nd Ed.*
Carol Chaitkin, M.S., and Lynn Gore, M.A.
ISBN 978-0-7641-4762-3, $9.99, *Can$11.50*

Painless Geometry, 2nd Ed.
Lynette Long, Ph.D.
ISBN 978-0-7641-4230-7, $9.99, *Can$11.99*

Painless Grammar, 3rd Ed.*
Rebecca S. Elliott, Ph.D.
ISBN 978-0-7641-4712-8, $9.99, *Can$11.99*

Painless Italian, 2nd Ed.*
Marcel Danesi, Ph.D.
ISBN 978-0-7641-4761-6, $9.99, *Can$11.50*

Painless Math Word Problems, 2nd Ed.
Marcie Abramson, B.S., Ed.M.
ISBN 978-0-7641-4335-9, $9.99, *Can$11.99*

——To order——
Available at your local book store
or visit **www.barronseduc.com**

Painless Poetry, 2nd Ed.
Mary Elizabeth
ISBN 978-0-7641-4591-9, $9.99, *Can$11.99*

Painless Pre-Algebra
Amy Stahl
ISBN 978-0-7641-4588-9, $9.99, *Can$11.99*

Painless Reading Comprehension, 2nd Ed.*
Darolyn E. Jones, Ed.D.
ISBN 978-0-7641-4763-0, $9.99, *Can$11.50*

Painless Spanish, 2nd Ed.*
Carlos B. Vega
ISBN 978-0-7641-4711-1, $9.99, *Can$11.99*

Painless Speaking, 2nd Ed.*
Mary Elizabeth
ISBN 978-1-4380-0003-9, $9.99, *Can$11.50*

Painless Spelling, 3rd Ed.*
Mary Elizabeth
ISBN 978-0-7641-4713-5, $9.99, *Can$11.99*

Painless Study Techniques
Michael Greenberg
ISBN 978-0-7641-4059-4, $9.99, *Can$11.99*

Painless Vocabulary, 2nd Ed.*
Michael Greenberg
ISBN 978-0-7641-4714-2, $9.99, *Can$11.99*

Painless Writing, 2nd Ed.
Jeffrey Strausser
ISBN 978-0-7641-4234-5, $9.99, *Can$11.99*

Barron's Educational Series, Inc.
250 Wireless Blvd.
Hauppauge, N.Y. 11788
Order toll-free: 1-800-645-3476
In Canada:
Georgetown Book Warehouse
34 Armstrong Ave.
Georgetown, Ontario L7G 4R9
Canadian orders: 1-800-247-7160

Prices subject to change without notice.

(#79) R5/12

MAXIMIZE YOUR MATH SKILLS!

E-Z ALGEBRA
5th Edition

Douglas Downing, Ph.D.

Topics covered in this detailed review of algebra include general rules for dealing with numbers, equations, negative numbers and integers, fractions and rational numbers, exponents, roots and real numbers, algebraic expressions, functions, graphs, systems of two equations, quadratic equations, circles, ellipses, parabolas, polynomials, numerical series, permutations, combinations, the binomial formula, proofs by mathematical induction, exponential functions and logarithms, simultaneous equations and matrices, and imaginary numbers. Exercises follow each chapter with answers at the end of the book.

(978-0-7641-4257-4) $16.99, *Can$19.99*

E-Z ARITHMETIC
5th Edition

Edward Williams and Katie Prindle

A brush-up for students and general readers, this book reviews all arithmetic operations—addition, subtraction, multiplication, and division, and calculations with fractions, decimals, and percentages. Includes practice exercises with answers.

(978-0-7641-4466-0) $16.99, *Can$19.99*

E-Z MATH
5th Edition

Anthony Prindle and Katie Prindle

Barron's *E-Z Math* reviews whole numbers, fractions, percentages, algebra, geometry, trigonometry, word problems, probability, and statistics. The book also presents a diagnostic and a practice test with answers.

(978-0-7641-4132-4) $14.99, *Can$17.99*

MATH WORD PROBLEMS THE EASY WAY

David Ebner

Readers start with exercises in algebra and progress to trigonometry and calculus. Each word problem breaks down into four successive parts: statement of the problem; its analysis; a work area; and the answer. An appendix presents solutions to all exercises and tests plus trigonometric tables.

(978-0-7641-1871-5) $16.99, *Can$19.99*

BARRON'S MATHEMATICS STUDY DICTIONARY

Frank Tapson, with consulting author, Robert A. Atkins

A valuable homework helper and classroom supplement for middle school and high school students. Presented here are more than 1,000 math definitions, along with worked-out examples. Illustrative charts and diagrams cover everything from geometry to statistics to charting vectors.

(978-0-7641-0303-2) $14.99
Not Available in Canada

Barron's Educational Series, Inc.
250 Wireless Blvd.
Hauppauge, N.Y. 11788
Order toll-free: 1-800-645-3476

In Canada:
Georgetown Book Warehouse
34 Armstrong Ave.
Georgetown, Ontario L7G 4R9
Canadian orders: 1-800-247-7160

Prices subject to change without notice.

— To order —
Available at your local book store
or visit **www.barronseduc.com**

(#142) R 2/11

When You Need to Brush Up on Your Math Barron's Has the Answer

Forgotten Algebra, 4th Edition

Barbara Lee Bleau, Ph.D.

This self-teaching brush-up course was written for students who need more math background before taking calculus, or who are preparing for a standardized exam such as the GRE or GMAT. Set up as a workbook, ***Forgotten Algebra*** reviews signed numbers, symbols, and first-degree equations, and then progresses to include logarithms and right triangles. Work units include examples, problems, and exercises with detailed solutions. Systematic presentation of subject matter contains all the algebraic information learners need for mastery of algebra.

Paperback, ISBN 978-1-4380-0150-0, $16.99, *Can$19.50*

Forgotten Calculus, 3rd Edition

Barbara Lee Bleau, Ph.D.

This combined textbook and workbook makes a good teach-yourself refresher course for those who took a calculus course in school, have since forgotten most of what they learned, and now need practical calculus for business purposes or advanced education. The book is also useful as a supplementary text for students taking calculus and finding it a struggle. Each progressive work unit offers clear instruction and worked-out examples. Emphasis has been placed on business and economic applications. Topics covered include functions and their graphs, derivatives, optimization problems, exponential and logarithmic functions, integration, and partial derivatives.

Paperback, ISBN 978-0-7641-1998-9, $16.95, *Can$19.99*

Prices subject to change without notice.

Available at your local book store or visit **www.barronseduc.com**

Barron's Educational Series, Inc.
250 Wireless Blvd. Hauppauge, NY 11788
Order toll-free: 1-800-645-3476

In Canada:
Georgetown Book Warehouse
34 Armstrong Ave.
Georgetown, Ontario L7G 4R9
Canadian orders: 1-800-247-7160

(#264) R4/13

Notes

Notes

Get access to your exclusive PAINLESS PRE-ALGEBRA mobile app

Just visit *www.BarronsBooks.com/painless.html*

To face the ultimate Pre-Algebra and Arcade Action Game Challenge!

A special **FREE** bonus

for every student who purchases *Painless Pre-Algebra*

Barron's is taking *Painless Pre-Algebra* to the next level — ***FUN***!

Just visit www.barronsbooks.com/painless.html. There you will find instructions on how to access your FREE app that contains three quizzes to test your knowledge and three EXCLUSIVE arcade games to test your skill!